The Secrets of Soviet Cosmonauts

Maria Rosa Menzio

The Secrets of Soviet Cosmonauts

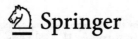

Maria Rosa Menzio
Associazione Culturale Teatro e Science
Turin, Italy

ISBN 978-3-031-09651-8   ISBN 978-3-031-09652-5   (eBook)
https://doi.org/10.1007/978-3-031-09652-5

Cover picture: © Angela Betta

This Springer imprint is published by the registered company Springer Nature Switzerland AG
The registered company address is: Gewerbestrasse 11, 6330 Cham, Switzerland

Foreword

In the five chapters into which it is divided, Maria Rosa Menzio's book retraces the milestones of the space conquest, from the second half of the '50s, when the first artificial satellite was launched, up to our days with the International Space Station.

The story is fascinating, both for those who lived those years and for those who want to know how things went and, perhaps, remove some doubts. Therefore, we discover the fundamental figures: Krushchev, Kamanin and Korolev. In the Soviet Union of the time, they contributed to what was a sort of further "cold war" (still against the USA) fought to the sound of victories. Unfortunately, they missed out on the final one, the conquest of the Moon!

It is sad to know the end of the dog Laika, whose fate was already marked in the organization of the mission. It is also sad to know the details of the death, at only 34 years old, of Yuri Gagarin, the first man in space, one of Russia's national heroes. There are still unconfirmed details about how things went. The cause of the sudden nosedive to the ground of the MiG-15UTI piloted by Gagarin, together with Vladimir Serjogin, an expert test pilot, remains a mystery. All of this more than 40 years after the fact, and even after the declassification of numerous documents, which took place in part after the dissolution of the USSR and, more recently, even with the consent of President Vladimir Putin.

Through minimalist but effective prose, Maria Rosa Menzio reports testimonies and accounts from different sources. In some cases, these take their cue from the enormous amount of documents that have come to light many years after the fact, unfortunate for history, but to the benefit of the reader. There are, for example, the stories of Valentin Bondarenko and Grigory Neljubov, both promising Soviet cosmonauts, who were part of the small group of pilots selected with Gagarin for the first flight into space. For different reasons, the Soviet regime preferred to erase them, even their very existence. But there are also interesting technical-scientific curiosities about the various cosmodromes and the reason why launches still take place from Baikonur, located in Kazakistan, to which Russia pays a substantial annual rent.

As a woman, Maria Rosa Menzio focuses on the figure of Valentina Tereshkova and on the reasons why she was chosen for the first "female" mission. Other candidates seemed to have superior characteristics, knowledge and preparation: among those excluded without apparently valid reasons, the case of Marina Popovich, known as "Madame MiG", is particularly interesting. The exploits of all of the selected candidates are described and their strengths and weaknesses revealed. The 70 h and 50 min spent by Valentina Tereshkova inside Vostok 5 are analyzed both for what was known at the time and for what was discovered later.

There is a curious reconstruction, perhaps veering into fantasy, of the meeting between Nikolaj Kamanin, responsible for the training of Soviet cosmonauts, and John Glenn, the first American to orbit the Earth with the Mercury 6 capsule: during a barbecue organized in the garden of Glenn's house, Kamanin had the feeling that the Americans were about to launch a woman into space. Back in Moscow, the man convinced Krushchev to speed up the female program: the candidates were already there and it was only necessary to intensify the training and establish a plan that would lead to the launch of a woman in the shortest possible time, to beat the US.

Result: Valentina Tereshkova flew on June 16, 1963, while Sally Ride, the first American to reach space, did so on June 18, 1983: 20 years later!

But in between, there was the conquest of the Moon, and this is another story that, in Menzio's book, is colored with interest, at least for the general public. She makes known the tragicomic Soviet attempt to land on the Moon, a few minutes before the Americans, at least one automatic probe, potentially able to collect soil samples and return unscathed to Earth.

Moreover, the author—as the mathematician that she is—uses iron logic to tell the "Moon landing deniers" how only the USSR could have had valid information that could have disproved the US's claims, but, since they didn't.

Speaking of more recent times, we discover the undisputed technical characteristics of the Soviet shuttle, almost a carbon copy of the US Shuttle, called Buran. This one flew unmanned (and returned) only once, and then "rotted" in the remote hangars of the immense Baikonur cosmodrome, after, however, having been shown to the world in 1989 at the Paris-Le Bourget International Aeronautics and Space Show, transported [?] on the back of a huge Antonov An-225 Mriya.

Of course, a book titled "The Secrets of the Soviet Cosmonauts" could not fail to include news, curiosities and even inferences about the incidents known as the "Nedelin Catastrophe" and the "Plesetsk Cosmodrome Tragedy," as well as the topic of cosmonauts lost in space: those who were perhaps sent into space unsuccessfully before Yuri Gagarin. On the other hand, securing new capsules for humans was always many steps behind the will of the politicians. Maria Rosa Menzio tells us how Gagarin himself had opposed the launch of what was rightly called the "flying coffin," the Soyuz 1 (two hundred and three documented construction defects), on board of which his friend Komarov died.

Finally, we talk about the ISS and the fruits of collaboration between Russia, the US and the rest of the West. In this regard, it is so hard to grasp how the world has changed from the time that the book was conceived and written to the time that it appears in its editorial guise. Is the future of the ISS and what it has represented for years really compromised?

April 2022

Attilio Ferrari
Professor Emeritus of Astrophysics
University of Turin
Turin, Italy

Acknowledgments

Without the Springer-Verlag publishing house, this book would never have existed: great credit is therefore due to Drs. Marina Forlizzi and Barbara Amorese and to Dr. Boopalan Renu.

Also, without the patient work of my editor Marc Beschler, the volume would never have seen the light of day.

I would also like to thank Anya Leonova, who checked and revised Chaps. 1, 2, 4 and 5 with great care and interest. Moreover, being Russian, she opened the way for me to access information that has never been made available in Italy.

My thanks also go to Jahid Chowdhury, for the patient revision of my imperfect English.

My gratitude goes out to the people who gave me permission to publish pieces of their intellectual property: Marco Zambianchi, James Oberg, Slava Gerovitch, Davide Lizzani and Emilio Cozzi, as well as the Russia Beyond website, mentioned in the text, that was a precious mine of information.

I may have forgotten someone and I apologize in advance.

The last and most important thanks go to my partner, Fulvio Cavallucci, tireless reviser of the various drafts.

Contents

1

From the Beginnings to 1960

Abstract After a quick survey of the major Soviet ventures into space, the "boundary conditions" that facilitated these successes are discussed. First, the three main figures who made it possible are outlined: Khrushchev, Kamanin and Korolev. An overview is then given as to the sort of homes in which the Soviets lived in those years, followed by a description of the Russian cosmodromes: from Baikonur, where many space enterprises had their genesis, to its successor Vostochny, in the far east of Russia, up to the floating cosmodrome. Finally, we talk about the story of the dog Laika, the first living being sent into orbit, the catastrophe brought about by Marshal Nedelin and the myth of the lost cosmonauts.

1.1 Introduction

> Without heroes, we are all ordinary people, and we
>
> do not know how far we can go. (Bernard Malamud)

In 1835, the memoir *On War* by Prussian general Carl von Clausewitz was published posthumously. We quote from it:

> War is but the continuation of politics by other means. It is not merely a political act, but a true instrument of politics, a continuation of the political procedure.

© The Author(s), under exclusive license to Springer Nature
Switzerland AG 2022
M. R. Menzio, *The Secrets of Soviet Cosmonauts*,
https://doi.org/10.1007/978-3-031-09652-5_1

We would add the following: the Cold War was also a continuation of politics. This included that aspect that represented the race to succeed in the area of achievements in space.

Let's take a look at the escalation that led the USSR to so many successes in orbit, even if the record in that regard ultimately went to the US.

The following are the various USSR spacecraft (somewhat uniformly named) that were launched into space.

In chronological order, we have Sputnik, Luna, Vostok, Voskhod, Venera, Soyuz, Cosmos, Saljut, Orion, Mars, Mir and Vega.

For two decades, the new arena of the Cold War between the US and the USSR was space. It was a fight without quarter in which the aim was to astound the world, both with technology and advertising. And in terms of espionage, neither adversary spared the other.

It all began on October 4, 1957, when Sputnik 1, the first artificial satellite to fly around the Earth, was launched from the Baikonur cosmodrome—a considerable source of national pride—reaching a low elliptical orbit. Sputnik 1 sent signals from space, making the initial score 1 to 0 in favor of the USSR. The US was totally unprepared for the "beep" that was broadcast by the satellite for almost three weeks. For humankind, the "space age" had begun. Not even a month had passed before the USSR's score had been elevated to 2-0: on November 3, 1957, Sputnik 2 took off from Baikonur. Another great triumph! Additionally, there was a living being in the satellite, the dog Laika.

But Laika, despite making it into orbit alive, later died, evidently of asphyxia. The attempt to bring a living being back to earth from space was not successful until 1960, with the dogs Belka and Strelka.

Up to that time, the US had been watching in astonishment, but in 1957, they did their utmost to change the trajectory. Their president, Dwight Eisenhower, proclaimed the beginning of the Vanguard project: the launch of a satellite between 1957 and 1958 was announced. The US had been defeated chronologically, but was clearly determined not to lag too far behind, a commitment they attempted to fulfill by launching the first Vanguard rocket on December 6, 1957. However, it was a disaster: the rocket exploded on the launch pad.

The following year, in February, the attempt was repeated with the Explorer 1 rocket: it was jointly designed by the Army and the German engineer Wernher Von Braun, a former Nazi who had moved to the United States. Explorer 1 represented the US's first success. But the country then went further: on July 29, 1958, it founded its own space agency, calling on none other than Von Braun to direct it.[1] From there, the competition went on. On April 12, 1961, Yuri Gagarin was launched into space in Vostok I. This made it 3 to 0 in favor of the USSR. It was a memorable event, and it had the mark of the Soviets all over it.

A few days later, on May 5, 1961, it was the turn of Alan Shepard, the first American in space, who was launched aboard a Mercury 3 rocket, in a suborbital flight (i.e., not performing a full revolution). (The first American to make a complete lap above the atmosphere would be John Glenn, launched within a Mercury 6, an event that was to take place on February 20, 1962.)

But then the really big challenge emerged: on May 25, 1961, US President John Fitzgerald Kennedy, Eisenhower's successor, declared to Congress that the Apollo Program was to begin, with the ambition of getting a human being on the Moon within 10 years at most. It started with the Gemini program, the mission of which was to verify the technical feasibility.

But the next record to be set still had the USSR brand on it. On June 16, 1963, Valentina Tereskhova became the first woman cosmonaut. The score was now 4 to 0.

The US response was not long in coming. On November 28, 1964, the US launched Mariner 4, the first probe to Mars, which successfully reached the planet in 1965. Score: 4–1.

This was followed by another great Soviet feat. On March 18, 1965, Soviet Alexej Leonov made the first spacewalk. Score: 5–1.

The US-USSR space race (or rather the run-up to it) just kept going. Indeed, the whole world was breathless on January 27, 1967, when the attempted launch of Apollo 1 resulted in it subsequently exploding on the launch pad. Despite this setback, the Apollo program continued.

Apollo 8 entered lunar orbit on Christmas 1968. In the same year, the USSR, which did not want to be left behind, launched the Zond 5 capsule, which contained the first animals to be sent into lunar orbit: two turtles.

By this time, it was clear that the US space program, although lacking in truly striking results, was a little bit ahead of the USSR.

The final victory that crushed all previous achievements took place on July 20, 1969, when Neil Armstrong, an American, set foot on the moon, the first human being to do so. The conquest of the moon was the pinnacle of the US space program. The Soviets did not want to declare themselves beaten: it was not widely known until later, but they had also tried, within the same days, to send a shuttle to the Moon. In the wake of that disappointment, they started

---

[1] When the Red Army arrived in Germany, von Braun boarded a train with false documents, and led 500 people through war-torn Germany to give themselves up to the Americans. The Americans realised they were dealing with top engineers and hired Von Braun.

working on Salyut, the first space station, which they launched on April 19, 1971.

But by that time, space competition had had its day: it was time to talk about cooperation. In July 1975, the conclusion of the Space Cold War was symbolized by the docking of Apollo 18 and Soyuz 19 in the first joint US-USSR space mission.

In the USSR, before each launch, the cosmonaut heard the Russian national anthem playing, and dedicated the mission to the motherland:

Russia - our sacred country,
Russia - our beloved land.
A powerful will, a great fame.
Are your heritage for all time.
Be glorious, our free homeland,
Eternal union of brotherly peoples,
Wisdom inherited from our ancestors!
Be glorious, homeland, we are proud of you!
From the Southern Seas to the Polar Circle.
Stretch our forests and fields.
You are unique in the world, you are inimitable,
Native land protected by God.
Wide spaces for dreams and for life.
Are open before us for the years to come.
Our fidelity to the Homeland gives us strength.
So it was, so it is and so it will always be!

The history of Russian cosmonautics stands as a novel dense with mysteries.

A lot of information has been classified for decades, even until the death of the protagonists. In the beginning, these events were told in a triumphant way, for the benefit of chronicles of the time.

After many years, some secrets have been revealed. It can now be seen that reality often exceeded fantasy.

Discovering what is hidden behind so many secrets is not easy.

The versions collected are often contradictory. Perhaps the whole truth is yet to be revealed, but this can help us to better understand the Russian soul.

1.2   The Three K's in the Lives of Soviet Cosmonauts

Who were the "founding fathers" of Soviet cosmonautics, from the political, preparatory and technical points of view? There are three names that stand out among the rest: Khrushchev, Kamanin and Korolev.

1.2.1   Nikita Sergeyevich Khrushchev

One cannot truly know the nature and character of a man

until one sees him wielding power. (Sophocles)

Khrushchev was the man who favored the establishment of the USSR as a power in space from a political perspective.

There are many interesting things to recall about this particular leader of the Russian Communist Party.

Regarding Stalin, Khrushchev significantly altered his perceived image by saying:

We are interested in how the cult of Stalin's person has been continually growing and has become, at a given moment, the source of a whole series of very serious deviations from party principles, party democracy and revolutionary legality.

In 1960, Khrushchev broke with the People's Republic of China, but even more importantly, he made a sensational gesture during a UN session. On October 12 of that year, there was a series of hot exchanges between the USSR delegation and the western delegations during the UN session. It was during the speech of British Prime Minister Morris Macmillan, who threw some sharp daggers in the USSR's direction, that Khrushchev **took his shoe off and started slamming it repeatedly on the table**.

In 1961, Khrushchev approved the plan for the construction of the Berlin Wall proposed by Walter Ulbricht, leader of the German Democratic Republic, in order to stop illegal emigration towards the West, which had become continuous and a source of serious concern.

But it was, above all, his idea of mass industrial construction that determined many Soviet destinies. Architectural form was to bend to the needs of the time, and Soviet architects no longer sought new forms, but rather got rid of decorative elements altogether, rationalizing the social function of architecture.

1.2.2 Nikolai Petrovich Kamanin

Nikolai Kamanin, the man who trained the cosmonauts—men and women—was, by certain records, born in 1908 and died in 1982.

However, according to his son Lev, Nikolai was actually born in 1909, in other words, a year had been added to his age. As it turns out, he had falsified his birth date in order to be chosen as a volunteer aviator in 1927. He subsequently passed the selection to become a pilot and finished aviator school in 1929. He couldn't have asked for a better instructor than the one he got: the legendary Soviet hero Victor Kholzunov.

As a pilot, Kamanin was sent near the railway that marks the border between Russia and China. He joined the legendary Lenin Regiment, the first Soviet airborne unit.

In 1934, Kamanin was awarded the newly established title of Hero of the Soviet Union. His heroic feat had been the rescue of the crew of the ship Chelyuskin, which had become trapped in the ice of the Chukchi Sea. The ship was then near Koljučin Island, between Russia and Alaska. The crew had managed to escape the wreckage by travelling along the ice floe and had been able to build an emergency airstrip. It had done so using only a few spades, snow shovels and two crowbars. A colossal undertaking, especially because the runway had to be rebuilt thirteen times before help arrived! Finally, in April 1934, the crew was airlifted to the village of Vankarem by seven brave pilots, one of whom was Kamanin.

In 1942, Kamanin led the fighter planes assigned to the defense of Stalingrad, and he showed great audacity. At the head of various war flocks, he participated in the battle of Kursk and led some missions into Nazi Germany.

After the Second World War, he worked to prepare pilots to fly the newly designed jet aircraft. Later, in 1960, he formed the first training school for cosmonauts. It was also he who chose Yuri Gagarin as the first man to fly in space.

In 1928, his son, Arkady Kamanin, was born. He too had a passion for flying: he spent the first years of his life in the airfields. He became famous for his uncanny knowledge about aircrafts, to the point that, just by hearing the engine "singing," he could determine what kind of plane it was.

Arkady soon became interested in airplanes from a practical point of view as well: in fact, he started to work as a laborer in one of the plants in Moscow. In 1943, he arrived at Andreapol airfield, where his father's division was based. The documents tell us that the boy was enlisted at the age of fourteen years, and learned the various techniques of flight on two-seater aircrafts piloted by soldiers, who often made him fly the planes in their stead.

It seems that once, during an exercise, Arkady had to land a Plikarpov PO-2 all by himself, because the pilot was injured.

The boy passed all the official exams to fly the PO-2 and, at only fifteen years old, became the youngest pilot in World War II. To celebrate, he was allowed to paint a lightning bolt on the fuselage of his plane. He had several medals, but he was never permitted to become famous: fate had arranged otherwise.

After the end of the war, Arkady's idea was to enter the Academy of Military Aeronautical Engineering in Zhukovsky. He would have surely become an aviation star.

Sadly, a case of fulminating meningitis killed him in 1947.

He was only nineteen years old.

Perhaps it was, in part, to compensate for the loss and the existential void left by his son Arkady that Kamanin later treated Gagarin as if he were his own son.

1.2.3 Sergei Pavlovič Korolev

Sergei Korolev was the man who designed rockets and spacecraft for the Soviet space program.

Korolev was a brilliant aircraft designer. His diploma project ("The glider sk-4") was approved for construction even before Korolev had had a chance to present it officially. Evidence suggests that the first three gliders he designed all went into production and were highly lauded. The story goes that, in 1929, Korolev read Tsiolkovsky's "The exploration of cosmic space by means of reaction devices" and, from there on, was fascinated by ideas around space exploration.[2]

The wonderful adventures of human beings in space began thanks to the work of this brilliant engineer, who was later to suffer great injustice. Born in 1907 in Ukraine, he grew up alone after his parents split up. Once he had obtained a degree in aeronautical engineering from Moscow Polytechnic, he joined the group working on the TB-3 bomber, after which, in 1930, he experimented with Russia's first small liquid-propellant rocket and published the paper "*Rocket Flight in the Stratosphere*".

But at the age of thirty-one, Korolev became a victim of the Stalinist "purges", arrested in 1938, slapped with utterly spurious charges and sentenced to hard labor in gold mines. Thus, early in his life, the man who

---

[2] Konstantin Tsiolkovski is revered in Russia as a genius who foresaw humanity's exploration of the cosmos. His ideas were instrumental in envisioning said exploration.

would lead the Soviet Union to triumph in its space adventures was subject to forced labor.

Korolev was initially sentenced to 15 years of labor work in the Kolyma and Maldyak gold mines. However, through the efforts of his mother and others, he was returned to Moscow in 1940, retried and subsequently had his sentence cut to 8 years. At this time, he was placed in the same sharashka[3] in Moscow where Andrei Tupolev, the aeronautical engineer, was being held, and was then later moved to another sharashka in Kazan. Tupolev had already been released by then. He knew Korolev from the Bauman Institute— Tupolev had been Korolev's diploma adviser. Rumor has it that Tupolev may have asked the "administration" for Korolev to be placed in his sharashka. Korolev still had several years of imprisonment left to serve, but he was young and worked in these sharashkas with excellent results. Everything the brilliant engineer applied himself to was transferable to the military field, and thus Stalin became interested in his experiments and ordered his release in 1944, although Korolev's name was never officially rehabilitated in those years.

At the same time that Korolev was working in Siberia in conditions of semi-slavery, what was happening in the US? Robert Goddard, the father of American rocketry, was receiving more substantial research funding from the wealthy Guggenheim family.

By the end of World War II, Nazi Germany had designed the V-2 ballistic missile. In 1946, Stalin designated Korolev as chief designer of the Special Design Bureau, asking him to devise a rocket similar to the V-2, but more powerful, i.e., with longer range.

This rocket was Korolev's big break. While the work was in progress, as a counterpart, he asked the government to support him in a project to send the first artificial satellite into orbit around the Earth. The challenge was accepted: he became the "*von Braun of the Kremlin*."

This is how the rockets called "Semyorka" were born. As the work progressed, they became more and more powerful, until, in 1957, Sputnik 1, the first artificial satellite, was launched. The event was of immense historical importance, establishing the USSR as the first power in the branch of space exploration. But there was no glory for Korolev: Stalin forced him to live under a false identity, and his name was kept a military secret. Misunderstood and opposed by the military and Soviet hierarchs, Korolev then devoted himself completely to the conquest of the Moon. A variant of the Semyorka-7 rocket eventually did reach Earth's satellite and photographed its hidden face. Korolev then designed the gigantic N1 rocket, also called Hercules N1. The

---

[3] A type of labor camp that was actually a secret study center.

N1, which had its own lander, was supposed to allow one person to land on the Moon while another remained in orbit waiting for him. But it was not successful, as we will see later.

Korolev could not even apply for the Nobel Prize: twice, after the launch of Sputnik 1, the Nobel committee asked the Soviet government for the name of the chief designer, but Khrushchev replied:

> We cannot point to a single person, it is the entire people who are building the new technology.

For almost twenty years, Korolev's name, and even more so his role as '*Glavnyj Konstruktor*'—Chief Builder—of the USSR's space programme, was the Soviet government's best-kept secret. His identity was kept under wraps until his death; in fact, official recognition was given only later. As to the false accusation that put him in prison, he was exonerated of his "crime" during his lifetime: the sabotage charges for which he was imprisoned could not be proved—but the state offered neither an apology nor compensation.

However, the brilliant designer's problems did not end there. He had health problems, first experiencing serious heart distress, and then, in 1966, undergoing an operation for polyps in his intestine.

After the operation, he did not regain consciousness. Sergei Korolev died on January 14, 1966.

1.3   Khrushchev's Bricks

> I have a cooking corner, a living room corner and a sleeping corner. All in the same corner. (Boris Makaresko)

Before becoming famous, various cosmonauts lived in very poor housing, as did so many of their countryfolk.

Privileged Russians, writers and members of the Academy of Sciences, lived together. But what about the others? What did the various houses in the Soviet cities look like?

Hundreds of thousands of people lived in what were known as "kommunalki," that is, houses where several families lived together, with only one kitchen and one toilet.

How did the kommunalki come into being? After the Revolution of 1917, all housing was nationalized. The government began to settle new tenants in the apartments of the rich, in a rapid process of urbanization. In shared apartments, people lived as one big family, and this way of living succeeded in shaping many generations of citizens both politically and socially, the Soviets first, and the Russians later.

When people arrived in the big cities, they were lost; they had no urban habits and did not know where to live. But if they found a job in a factory or a state agency, then they could be assigned a room in a shared apartment, for an area of about ten square meters for each adult and five square meters for each child, although the values were subject to multiple changes over time. People who, until a few days earlier, had been peasants became roommates of pre-revolutionary intellectuals; cooks shared bathrooms with university professors. It was not easy to live like this, among different social classes, but everything corresponded to the official ideology, which did not permit class distinctions.

Let's take a closer look at what a kommunalka looked like. Georgy Manaev, in *Russia Beyond*, gave a full description of it.[4]

The entrance door to the apartment might have a separate bell for each tenant; oftentimes, it was a single bell, but the number of rings was different for each tenant. One of the walls at the entrance was home to a bevy of electricity meters; each tenant paid their own electricity bill.

There was no coat rack or common closet.

The space in the hallways was occupied by bicycles and skis.

There was only one telephone on the wall; occasionally, but exceedingly rarely, there would be private ones in the rooms.

There was a common entrance hall along which opened the doors to the individual rooms.

Each room had its own doormat in front of it.

Everyone kept their clothes, coats and shoes in their own room.

In the bathroom, there were many washing basins and many bars of soap. Sometimes, there was only a single washing basin, and tenants had to take turns using the bathroom. The toilet was separate from the bathroom. Many of these spaces had previously been single family apartments, which had been transformed into multifamily living spaces where every family/tenant occupied a single or, at most, two rooms, depending on the size of the family.

The tenants had to be careful to use their own soap and their own personal toilet seat as well.

In a popular joke about the kommunalki, a lady is washing herself in the communal bathroom. Suddenly, she notices that a neighbor is spying on her.

---

[4] https://it.rbth.com/societa/2014/10/24/cosi_si_viveva_al_tempo_delle_kommunalki_33165.

To the woman's scream, the neighbor replies, *"I'm not looking at you; I just wanted to see which soap you're washing with."*

As the scholar Ilja Utekhin writes in his book "Ocherki kommunalnogo byta" ("Bits of life in the kommunalki"), in the bathroom, you could find attached notices such as:

"Leave the bathroom clean after use."

"Do not throw paper in the toilet," or, more simply,

"Don't take other people's things."

These are notices that we are far more used to seeing in public places! But then, the spaces that the Russians lived in almost were, in a sense, public places.

And what about the cleaning of common spaces? Easy: all of the inhabitants of the apartment took turns cleaning. It was based on a calendar hung in the hallway, visible to all. Then, they would collect money for repairs to the electrical and plumbing systems and for other common needs. Anyone who didn't participate in the common cause would find everyone else in the apartment turned against them. This could make life hell for the unfortunate in the kommunalka.

They met mostly in the kitchen, making food or washing dishes. Meetings were held when:

– matters of common interest needed to be resolved, or
– the behavior of individual tenants who disturbed the peace or engaged in "bad" conduct needed to be discussed.

"My neighbors know who came to see me yesterday. And they would like to know who came the day before yesterday," sang rock singer Fedor Chistjakov in the song "Kommunalnye kvartiry."

People gossiped a lot, personal hatreds were known occasionally to develop between tenants, and it was not at all uncommon for individuals to listen secretly to each other's conversations. On this basis, life in the kommunalki became a breeding ground for grudges that lasted for years. Sometimes, people resorted to terrible means: hiding iron wire in the neighbor's soap, or secretly pouring washing powder into the boiling soup of their enemy.

Conflicts, however, were a rare phenomenon.

Usually, the apartment dwellers found a way to get along. It was much more comfortable and livable. And they even tried to help each other, to look after their neighbors' young children, to take turns caring for the elderly tenants, to help neighbors get a job, or to lend a hand to those in need. When the relationships between them were good, the neighbors could also tighten up a bit: for example, if there was a young couple with a child, and the husband came home from a long business trip, for one night, the neighbors could host the child in their room, so that the couple could be alone together.

In this way, cohabitation educated people both in social responsibility and in mutual help and support. Tenants from older generations could teach this to younger ones.

Irina Kagner, a veteran of a Moscow kommunalka, remembers:

> This is how both workers and intellectuals lived. Intellectuals were listened to, they were taken as examples. From them, one learned to live, to have taste. The children saw how they lived, and once they grew up, it could not be said that they had not received notions of culture.

In the late 1950s, mass construction began in the USSR. Finally, a lot of people would have access to a personal space in which to live. For many people, it was something incredible to have an individual apartment. Moscow resident Marina K. remembers:

> My grandfather and grandmother lived for many years in a kommunalka in Sretenka (Moscow district), where as many as forty people lived in addition to them. When they were finally assigned separate housing, my grandfather sat on the kitchen floor, leaning his back against the wall, and remained in that position for a long time, just to enjoy the silence.

The real boom of kommunalki-emptying began in the 1990s. At that time, businessmen wanted to buy the properties in the center of the two Russian capitals (Moscow and St. Petersburg). They were even prepared to offer an apartment to each of the old tenants from the kommunalki. Yet, not all kommunalki emptied out. In 2011, Moscow's Department of Housing Policy claimed that there were still 91,000 kommunalki in the Russian capital alone. Three years later, they accounted for two percent of all housing properties in Moscow.

But the demand for rooms in shared apartments is not decreasing at all. This is because not everyone can afford to rent an apartment in Moscow. Prices start from five hundred euros per month. If you opt for a room in a kommunalka, however, you can spend less: even as little as two hundred and fifty euros, a price that most young people and workers coming from other cities can afford. So, today, kommunalki continue to spring up. This is especially likely to happen when a married couple separates and divides the common property. Many Muscovites and Petersburgers rent out some rooms in their apartments, living off of "rent income," and thus creating a new generation of shared apartments.

But some aspects of Russian legislation also make it difficult to eliminate kommunalki. A family of four living in a kommunalka is entitled to a three-room apartment, but they must often settle for only a two-room apartment in exchange for the right to use the kommunalka, perhaps because it has the benefit of being located in the center of town. This is why the kommunalki are likely to last for a long time.

Let us now make a comparison between the kommunalki and the Italian houses of the Fanfani period. In Italy, the social housing plan tried to marry social Christianity to Marxist collectivism. It was an ideology that thought of the city based on its individual neighborhoods. It wanted housing that was not just a place to sleep. Both the interior and exterior were designed according to the plan, centralized distribution was called for, and there was much talk about the concept of the "minimum house" while stressing the importance of integrating housing and services. Then, there was the centrality of public space and the importance of meeting places. There was a real desire to create an idea of citizenship and not only of residence.

In Russia, after the kommunalki came the chruščёvki.

These were born out of the need to build housing in the shortest time and at the lowest cost possible. What were the categorical imperatives?

Here they are: from 1947 to 1951, Soviet architects were ordered to eliminate all non-essential elements.

Balconies and cellars? Superfluous! Eliminated from the projects.

Elevators? Superfluous! Also eliminated from the plans. This saved eight percent of the budget, but it also necessitated that the height of the buildings stop at the fourth floor.

The ceilings were lowered from 270 to 250 cm, which was still more than the 226 cm of Le Corbusier's "housing unit."

Construction was simplified to the bone.

This made it possible to assemble a prefab of panels in just fifteen days (some teams of workers managed it in a week). Interior finishes took another month.

The chruščёvki thus became the product (or the victim) of two factors:

– the mediocre quality of materials,
– the desire to build in as little time as possible.

For these reasons, people suffered from the cold in winter, the heat in summer, and were constantly subject to the noises and conversations of their neighbors. Not exactly the ideal situation for a population that came from the countryside, where the isbas were well protected and isolated. Ten years later, the production cycle of the series ended, but a lot of chruščëvki buildings are still there. Their gradual demolition started in the '90s, and still continues to this day.

In the Soviet apartments, size and floor plan were designed by architects. One family per dwelling, with eight personal square meters for each tenant. Again, the superfluous was excluded.

The architects started from the idea that a room could have a dual function: during the day, you ate and worked, at night, you slept.

But these limitations resulted in more apartments for everyone. The concept of personal space made its first appearance in the lives of the citizens.

One could leave the countryside or kommunalka and arrive in a new apartment, able to live in one's own way and shielded from the prying eyes of one's neighbors. One could host friends and talk to them about whatever one wanted: thus, the personal sphere of Soviet life was born.

Despite its shortcomings, the simply and sloppily built chruščëvka defeated the building crisis of the USSR. By the end of Khrushchev's rule, fifty-four million people had moved into the new apartments. After another five years, the number had risen to one hundred and twenty-seven million. The Soviet Union experienced macroscopic urbanization, and by 1961, the urban population exceeded the rural population.

The creation of mass housing produced a social revolution, which became one of the consequences of Khrushchev's thaw and led to the humanization of the relationship between the state and the citizen. This relationship was no longer based on the terror of repression, but on the desire to receive an apartment. In this way, the human being formed a personal environment, which was hardly compatible with the ideology of control.

1.4   All Cosmodromes of the USSR/Russia

In recent times, Russian cosmonautics, while waiting for (possible) future missions to the Moon or Mars, has mainly been engaged in projects conducted on the International Space Station (ISS), which can be reached by various means: the Soyuz spacecrafts, the Progress, Dragon and Cygnus shuttles, the H-II Transfer Vehicle (Japanese cargo shuttle) and, until 2015, the ATV (Automated Transfer Vehicle) developed by the European Space Agency.

The Russian Soyuz spacecrafts are the only ones able to carry cosmonauts and astronauts, as well as depart from the Baikonur cosmodrome, which, however, is not located in Russia. Within the Russian territory, there are other cosmodromes or military bases structured for the launch of ballistic missiles: Kapustin Yar, Yasny and Svobodny, in addition to Baikonur, Vostochny and Plesetsk, of which we will speak later, since they are the most important.

In the oblast[5] of Astrakhan, practically right on the Caspian Sea, there is Kapustin Yar, built in 1946 and considered by many as a military training field, although ballistic missile tests are carried out there, with warheads being launched into space.

In the Orenburg oblast, on the other hand, there is Yasny, a cosmodrome in use since 2006, so it is still quite new. Considered as a launch base for rockets, it is mainly used to bring satellites into low Earth orbit.

The cosmodrome located further east is Svobodny, in the Amur oblast, operational since 1996 and structured for the launch of light and medium class missiles. Its construction began because Baikonur, while technically a part of the Russian Federation, was actually within the borders of Kazakhstan, and the leaders of the Russian space program decided that the state needed its own cosmodrome at home. Svobodny is hardly ever actively used, thanks to the new facility at Vostochny.

Russia Beyond[6] gives us another piece of interesting news: Russia owns the only floating cosmodrome in the world. But it is more a source of trouble than satisfaction.

It's called the "Morskoj Start/Sea Launch," this cosmodrome that travels on a marine platform. In theory, it could be located at the height of the equator, where, thanks to the energy of the Earth's rotation, larger masses could theoretically be launched. So far, however, it has only been known for scandals, economic problems and corruption.

When it was bought by a Russian company, the floating cosmodrome raised many hopes. It was the only private cosmodrome, built for "heavy" private astronautics, and moreover, Russia had only been able to obtain it by overcoming many difficulties. In the whole world, no one else had a floating spaceport.

For these reasons, "Morskoj Start" made people dream. Elon Musk[7] himself has said that he envies it.

---

[5] A territorial subdivision used in some Slavic states and former Soviet republics. It corresponds roughly to a region or a province.

[6] See https://it.rbth.com/scienza-e-tech/84952-la-russia-possiede-lunico-cosmodromo of September 16, 2020.

[7] CEO of Space X and Tesla.

In the beginning, the project was international, despite having Russian "roots" since 1993. The Russians only managed to get hold of 25% of the shares. 15% of the rest were Ukrainian, 40% American and 20% Norwegian.

It all began with a self-propelled drilling platform, made by the Japanese: this was the "Odyssey", built in 1982 and transferred to a port on the Baltic Sea, after a serious accident in 1988 had engulfed it in flames off the coast of Great Britain. Before then, disused, her carcass had been moored for a long time in Dundee, a port in Scotland. It was subsequently towed and repaired in the shipyards of Vyborg, Russia.

The "Odyssey", with all its equipment attached, represented the sole structure of the "Morskoj Start" project. Since 1998, there have been 36 launches, 32 of which were successful. But this was too little to cover its costs, so, in 2009, it went into financial collapse. At first, the facility passed to Roscosmos, the Russian space agency, then, in 2016, it was acquired by the private space company S7 Space, which also includes Russia's second largest airline, S7.

It is estimated to have cost $100 million.

Comparisons have begun between S7 Space founder Vladislav Filjóv and Elon Musk. An interesting competition is expected. However, S7 did not buy something ready and profitable; on the contrary, the design work was a disaster, in spite of the enthusiasm.

The biggest difficulty was that it lacked rockets. After the Ukraine crisis of 2014, Ukrainian companies were no longer willing to provide Russia with the Zenit launch vehicle, adapted for sea launch. The problem continued even when the cosmodrome went private. No agreement could be reached between Ukraine, the United States and Russia.

> So the company relied on the Soyuz-5 rocket developed by Roscosmos, ready only in 2023. And since it is impossible to launch a Russian rocket from U.S. territory, it was necessary to transport the entire cosmodrome across the Pacific Ocean to the Littoral Territory (the Vladivostok region). This maneuver, in turn, required the creation of the entire coastal infrastructure from scratch. Throughout this time, the platform, commissioned in the 1990s, continued to age and remain inactive.

Thus, the last launch was in 2014. There were then other huge losses of money as a result of the pandemic. The private airline ran out of money and put the cosmodrome up for sale. But no one wanted to buy it, since it was not yielding any benefits, taking into account, among other things, competition from Elon Musk. Others said that, in the absence of buyers, S7 would demolish the "Morskoj Start."

In 2020, Russian Deputy Prime Minister Yury Borisov instead declared that he wanted to restore it, even at the cost of paying crazy amounts of money: he foresaw an expenditure of 35 billion rubles [even though the costs had previously been estimated at as much as 91 billion rubles, i.e., 8 times more than the purchase price of the project]. Borisov would obviously have decided this after a meeting with Putin. But no one is saying yet where the funding will come from. Borisov reiterated:

> It would be foolish of us not to restore the launch system from the sea and not to use it. All this is technically possible.

However, according to experts, the desire to restore the Sea Launch is only a matter of prestige, since the floating cosmodrome will compete with the Vostochnyj cosmodrome, and there will not be enough commissions for both projects. It is said that only a fool would invest in a project that has already suffered so many financial disasters and acquired such a very bad reputation.

1.5   The Baikonur and Vostochny Cosmodromes

Let's get a sense of the atmosphere of Baikonur through Claudia di Giorgio's article in *"La Repubblica.it", Culture & Science* of April 24, 2002, written on the occasion of the departure for space of the Italian astronaut Roberto Vittori:

> *Even before the take off, the presence of Roberto Vittori in Baikonur was already an event. It is the first time, in fact, that an Italian astronaut has become a cosmonaut, arriving in Kazakhstan as an official member of a Russian crew. A historical event, then, in a space base that, despite having been the protagonist of the major Soviet victories in the "space race" of the '50s and '60s, is actually very little known.*
>
> *The Baikonur Cosmodrome is the largest space base in the world. But in spite of its size (it covers an area almost as large as Abruzzo), for decades, it has not appeared on any geographical map and has been one of the most jealously guarded secrets of the Soviet Union. Even its name is a fake: the original Baikonur is a city 370 km away. But in 1955, when the Soviet authorities began building a new military rocket installation near a remote village of Kazakh shepherds, the center was named Baikonur in an attempt to confuse the Americans. The attempt failed: the CIA discovered the installations and photographed them from above as early as 1957, apparently thanks to a single railroad track that ran across the steppe in the direction of the base.*
>
> *Built at the height of the Cold War, Baikonur seems tailor-made to illustrate the expression "in the middle of nowhere." Deserted, greyish, improbably flat, the*

landscape surrounding the base lacks only craters to be a perfect copy of the Moon, the satellite that the Soviets have dreamed of and unsuccessfully tried to conquer from here. And the evocation of lunar landscapes is only one of the many symbols that contribute to making Baikonur a time capsule that, through the space epic, encompasses the entire parabola of the Soviet Union.

It was from Baikonur that Sputnik, the first man-made object to leave the Earth's atmosphere, left in 1957; from here, the dog Laika was launched; from here, on April 12, 1961, Yuri Gagarin took off for the first human flight into space. And two years later, the former factory worker Valentina Tereskhova was launched from Baikonur, winning the title of the first woman in space and, as a further slap to the US, remaining in orbit for more hours than all those achieved by American astronauts up to that point combined. But those successes, largely springing from the fact that Baikonur is the starting base for all Russian manned flights, including those to Mir, have alternated with defeats.

This can be seen from the moment of arrival, when one discovers that the plane - the only one that takes people inside the base - lands using a runway intended for the Buran, the Soviet equivalent of the Shuttle, a reusable spacecraft, the pride and joy of the Soviet Union, the construction of which was interrupted for lack of funds at the end of 1988, after a single test flight. The installations that were used to launch it still dominate the area, but not even the pride of those who show them to journalists, in tours involuntarily suffused with the tenor of those of a cemetery, can hide the humidity that drips everywhere and the rust that is devouring them.

Additionally, the signs of the neglect suffered by the cosmodrome in the nineties are clearly visible in the middle of the steppe, among uneven roads, disused tracks, buildings with broken roofs and the omnipresent rust that devours the carcasses of vehicles abandoned in the fields.

Since 1992, Baikonur has been part of the territory of the independent Republic of Kazakhstan, but Russia has maintained control over the base in exchange for "rent" of 115 million dollars per year. However, in the darkest period of the crisis, while the launches continued, albeit to a reduced extent, those who worked and lived there found themselves deprived of salaries, medicine, food and even drinking water. Devastated by poverty and recurring epidemics, Baikonur City, where most of the personnel live, has gone from 100,000 to 20,000 inhabitants.

In recent years, however, things have begun to change. Some buildings have been restored, new homes have been built, and the base has resumed full opera-tion. The excellence of Russian aerospace technology, which once terrified the West, is now relaunching Baikonur, both in the context of the partnership in the Inter-national Space Station, into which the skills accumulated through Mir have been poured, (the word "Mir" in modern Russian can be translated as "peace" or "world", depending on the context), and in the commercial exploitation of space, of which the Russians have been pioneers, with unscrupulous ideas such as sponsoring launches and selling "tourist tickets." It was ideas like these that brought private foreign companies, new investment and new capital to Baikonur. And new hope. Even the Buran, base technicians say, may one day fly again.

In 2010, Russian President Vladimir Putin authorized the construction of the new Vostochny launch base, Russia's newest and easternmost spaceport. Right away, it displayed a number of shortcomings. First, we note that Vostochny is located 6 degrees north of Baikonur, so the total payload mass that will be launched into space from the new Russian cosmodrome will be slightly less. But while this launch base will initially be used to reduce Russia's dependence on the expensive Baikonur Cosmodrome, it will later replace it altogether.

In this regard, on January 4, 2018, in the monthly periodical Focus, Davide Lizzani broke further news linking Baikonur to Vostochny:

Russian satellite missing owing to incorrect coordinates.[8]

That is, a human error caused the loss of the signal from a $45 million satellite: it had been given the coordinates of a different departure site.

Technicians thought that the satellite had departed from Baikonur, Kazakhstan, but instead, it departed from Vostochny spaceport in southeastern Russia... This (human) error was fatal for the satellite launched by the Russian space agency Roscosmos.

The Meteor-M, which was supposed to acquire meteorological information, is lost in space, with no chance of recovery.

After losing connection with the rocket, technicians at the base first thought it might be a problem with the second stage (the second part of the launcher rocket, which accelerates the satellite to orbital speed), but later, the true problem became clear. "The rocket was scheduled to start from Baikonur," Russian Vice President Dimitry Rogozin revealed to state TV. After the preliminary reconstruction of the disaster, in fact, the subsequent investigation confirmed this unfortunate hypothesis.

Starting on the wrong foot, the rocket had not followed the trajectory that the flight engineers expected, and this had made communication with the rocket, and therefore any attempt to adjust its flight parameters, impossible. Now, Meteor-M is missing, together with 18 other microsatellites of American, Japanese, Canadian and German origin. It is not a good start for the new Russian spaceport, which, at a cost of seven billion dollars, is supposed to be a state-of-the-art launch base.

Nonetheless, today, it is still from Baikonur that astronauts leave for the International Space Station.

As to the new Russian cosmodrome of Vostochny, there is more to be reported.

The first launch, promised for 2015, with the new cosmodrome still under construction, was postponed to 2016, theoretically on April 12, the anniversary of the historic launch into orbit of Yuri Gagarin.

---

[8] See https://www.focus.it/scienza/spazio/satellite-russo-disperso.

To make things easier, it was decided that the Soyuz rocket would be used as a carrier, an old and consolidated technology, consisting of three parts: the orbital module, the re-entry capsule and the service module. The programmed date was quite flagrantly missed, because—in a turn almost completely unheard of in Soviet astronautics—90s before departure, the countdown had to be interrupted. The problem was in the control system of the rocket, purportedly a defective cable. To think that, in order to be present at the inauguration, Putin had flown 5.500 km from Moscow!

Nice job, especially for a cosmodrome that does not have that many successes to crow about in the first place!

The launch was re-organized for the following day.

President Putin did not hesitate to give a good scolding to Roscosmos's top management, reiterating in every way that the fault for the failure was that of the rocket, and not of the cosmodrome.

With the launch finally achieved, three test satellites were put into orbit: the Aist-2D, the Mikhailo Lomonosov and the SamSat-218.

But the new Vostochny cosmodrome is supposed to be the emblem of Russian technology, of an ultra-modern and flourishing nation, both in abstract and, especially, in applied sciences.

Russian plans called for the first manned launch from Vostochny in 2018. There were, however, major difficulties, none of which cared at all about Putin's schedule. But no official announcement had been made to that effect. The Russian Space Agency Roscosmos therefore had a big issue to deal with: Putin didn't even want to think about a postponement; he pounded and pounded on that nail and ordered that the first manned launch from Vostochny be done, without fail, in 2018. Deputy Prime Minister Rogozin, speaking in Star City, also said:

> I would like to set an important date, which no one will be able to doubt. I intend for the first manned launch from the Vostochny cosmodrome to take place in 2018.

In addition, President Vladimir Putin, who is never shy about letting his wishes be known, announced,

> We Russians will go to the Moon by 2030.

But there was another problem: the PTK-NP capsule, the associated Angara 5 launcher, all the logistics infrastructure and the launch pad were not ready until 2020. In order to please the Kremlin, the Space Agency worked hard and went back to Soyuz, replicating all the facilities at Baikonur from 2016 to

2020, only to decommission them after a very short time. In addition, all the recovery teams have been accustomed, for over 50 years, to working in the flat steppes of Kazakhstan, specifically in Baikonur. Instead, in the case of a failed landing, they would find themselves in a completely different environment, the taiga and the Siberian mountains, and meanwhile, new naval units would have to deal with the Pacific Ocean.

But Roscosmos has already "tested the ground". In fact, some recovery teams have already done actual tests in the Pacific with Soyuz capsules. And mountain training for cosmonauts has been restored after a long period of abandonment. It is just that the cosmodrome that Putin wanted as a symbol of Russian pride has turned out to be very expensive. Moreover, according to Russian prosecutors, at least $165 million was embezzled during the construction process (while critics say that these numbers are grossly understated).

Regarding other failures, on June 19, 2019, Giuseppe Agliastro titled his article in *La Stampa* thusly:

Putin still seeks the conquest of space, but his Cosmodrome is a flop.

The first launch of a Soyuz rocket with 10 microsatellites had been set for May 2018, but was then postponed until the end of the year. It was subsequently carried out on December 27, 2018, using a Soyuz 2.1a. The primary payloads were two Russian government Earth observation satellites, Kanopus-V 5 and 6. Also on board were 26 small satellites that were deployed as secondary payloads. The launch of these small satellites was organized by GK Launch Services, a commercial subsidiary of Roscosmos. Despite the many enthusiastic news articles—Russian and otherwise—of launches occurring from Vostochny, even two in one year, the reality has been discouraging. In fact, the next launch, in 2019, was not successful:

...and apparently ended with a literal hole in the water: in the waters of the Atlantic to be precise, where - according to some sources cited by Russian agencies - the 19 satellites that had taken off a few hours earlier with a Soyuz carrier rocket would have miserably sunk.

Here's how it went down.

The Soyuz carrier rocket took off and was filmed live on state TV. Everything seemed to be going according to plan. Vice Premier Rogozin even congratulated the Russian space agency Roscosmos for *"giving the whole country a party"*. *"Glory to Russia!"* he exulted.

A couple of hours later, the first alarm was triggered: the most important of the 19 satellites sent into space, namely, the meteorological "sputnik" Meteor-M, had not sent the expected data after undocking from the rocket. Shortly afterwards, Roscosmos was forced to admit that the satellites - some of which belonged to organisations and companies from Canada, Japan, the USA and Germany - had not reached the expected orbit. Finally, the coup de grace: some sources reveal that there had been an error in calculating the trajectory of the Soyuz rocket and that of the Fregat system that was supposed to send the satellites into orbit. Conclusion (unofficial): it was likely that all of the satellites had crashed in the Atlantic, near Antarctica.

So many difficulties, and so many scandals, investigations and arrests of certain corrupt managers, but Russia is still betting on Vostochny. Not only because, in this way, it will be possible to free itself from the old cosmodrome of Baikonur and from the astronomical (an entirely appropriate word) rent paid to Kazakhstan every year, but also because Putin wants to be remembered as the man who reinvigorated the space sector.

But the "unlucky" launches are not the Vostochny's only problem: among corruption, delays in work and hunger strikes by unpaid workers, the cosmodrome desired by Putin in the Russian Far East seems far from what it should be: the flagship of a renewed Russian space sector with great ambitions.

Despite this, Putin is still counting on carrying out the following from Vostochny:

– in 2022, the first flight with cosmonauts
– by 2030, to go to the Moon
– the next destination will be Mars, on a date yet to be determined.

Beautiful goals, but the economic crisis, with the addition of Covid, will force Russia to cancel a great deal of the space sector's funding.

1.6   Laika, Woof Woof in Space

Dogs have but one defect: they believe men. (Elian J. Finbert)

Commander-in-Chief of the U.S. Strategic Air Command General Thomas Power, supporting the allocation of funds to the US space program, declared:

Whoever is first to make their claim in space will own it. And we simply cannot allow ourselves to lose this contest for supremacy in space.

The conquest of space, the race to be first in this achievement or that between the US and the USSR, would inevitably make a brief pass through the launch of an animal. The question that the space directors asked themselves was: what animal to send into space? The Russians preferred dogs to mice or monkeys: they lent themselves more to the image of the "hero." Vladimir Ponomarenko, who was then head of the Soviet space academy, said:

Monkeys are always misbehaving, they try to touch everything they can, whereas the dog is a friend of man, and is easy to train.

The poor monkeys were thus discarded, with a note of blame for their exclusion. The dogs, on the other hand, behaved well. And for that, they paid a price.

Her name was Laika, the little dog that was launched into the sky in November 1957. Her name means "Little barker" in Russian. She was sent into space with the Soviet capsule Sputnik 2.

Her destiny was certain death, because the capsule was not designed for the return of the animal to earth. Laika's tragic end was marked from the beginning of the mission. This despite the fact that she had "behaved well."

In space, they were able to measure her blood pressure, heart rate and respiratory rate. In other words, all vital signs.

But according to some sources, Laika died only 67 min after launch. Sixty-seven.

In reality, the dog was named Kudrjavka, "Curly." In addition, the English called her Muttnik (a cross between "mutt" and the name of the Sputnik capsule). So why did she become known as Laika?

The dog would become a star named Laika as the result of a misunderstanding. A Western journalist asked a mission manager what the dog's name was. The woman being interviewed thought that they wanted to know what breed it was. So, she answered "Laika."

Laika are Siberian dogs similar to huskies. The breed was chosen because they are very resistant to extreme conditions, especially to low temperatures. And yet, in a mockery of fate, Laika died (according to some sources) from dehydration because of the excessive heat…

But how was the mission designed?

Sputnik 1, the first satellite in space, had been a success. To enhance the Russian record, more were needed. More satellites and more successes. Moscow's ultimate goal was to launch humans.

Two satellites, also of the Sputnik type, were on their way to being finished, but neither would be by November 7, 1957. What a jinx! It would have dashed the Soviet dream of sending up a satellite carrying a human being on the perfect day: the fortieth anniversary of the end of the October Revolution.

But, aspirations aside, the idea was a complicated one. Neither the Soviet Union nor the United States of America had any experience sending a human being into space. What would happen to them? It was a mystery, about which no one had the faintest idea. It was not even known whether the human body could survive for long in weightlessness.

So, some data had to be collected before any launch with a human crew on board could be considered. There had to be some other way to get it.

What other way? Easy: send an animal.

A beast could be sacrificed. It could be made to go down in history as a simple soldier, and the officers—the humans—would come later.

Thus, the decision was made to start construction of a fourth satellite, a much simpler one, to be launched by November 7.

Everything was top secret. Absolute silence had to be maintained on the decision to send a dog into space. Maybe it was their turn because they were small animals.

Officially, Laika was a three-year-old stray dog found by chance in Moscow. She was a mixed breed, half Husky and half Terrier. Why her? That information is still top secret.

Three dogs were selected for the Sputnik missions: Laika, Albina and Mushka. Albina was Laika's backup, while Mushka was used to test the vital systems in the capsule. All three underwent intensive—read: hellish—training, directed by Oleg Gazenko, the program manager, the one who had chosen Laika for the first space flight.

During the training, the animals had to get used to small spaces: they remained in very small cages for up to twenty days in a row. They suffered a lot, but it was all right. After all, they were just animals.

Not entirely meek animals, however: in fact, Laika began to become increasingly nervous and, for a while, the training had to be suspended. Damn it! When the training was resumed, the animals were forced to stay in centrifuges, inside of which the vibrations and noises of the launch were reproduced. The animals' pulses beat twice as fast as normal. The dogs were terrified, they suffered a lot and Laika herself began to have panic attacks and exhibit rage.

Unofficial sources claim that the dog was brought on board the satellite three days before the launch, looked after by two technicians, to give her "*time to get used to it!*"

It was cold in space, and the capsule had to be connected to a heating system to maintain a constant temperature. In addition, before the launch, some electrodes would be attached to the animal's body to transmit vital signs to the control centre, namely, heartbeat, pressure and respiration.

The capsule was a cone with three iron trunks inside it, one of which contained Laika. The cone was placed on a Semyorka R-7I rocket. The capsule weighed a total of eighteen pounds, to which the six pounds of the animal had to be added. The inside of the satellite was lined and the space was large enough for Laika to lie down or stand up. As she wished. Such comfort! The temperature inside was 15 degrees and food and water had been prepared in the form of gel.

Laika was launched into space on November 3, 1957, at 2:30 am, from the Baikonur Cosmodrome. Her pulse was highly accelerated, and returned to normal only with the reduction of gravity. Some other Soviet sources claimed that signals continued to be received for a full seven hours. After that, absolute silence. But the official version from the government is that Laika survived for "more than four days."

Laika had not chosen to be a cosmonaut. She didn't know what she was getting into. Above all, she had no say in the matter.

The satellite re-entered the atmosphere five months later, on April 14, 1958. It had completed two thousand five hundred and seventy laps around the Earth and was completely destroyed. Re-entry into Earth's orbit was not possible, because the capsule was not equipped with a heat shield. So, as we said, the fate of Sputnik 2 was already marked at the time of launch. Laika had to die.

According to the government, from a technical point of view, Sputnik 2 was a success.

The launch of Laika into the cosmos disoriented the West.

The whole world now realized how advanced the Soviet Union was. In addition, compared to the US, it had a great advantage in the construction of satellites, and especially of launchers, which had longer ranges and greater load capacity.

Sputnik 2 was built in record time. They took the Semyorka carrier rocket from Sputnik 1 and modified it so as to send Sputnik 2 into an even higher orbit. Over the years, the Semyorka was further elaborated: they managed to construct a rocket with a range of 12,000 km and a payload of up to 5,370 kg. Astonishing.

Everyone understood that the Soviet Union had the means and the technology to deliver nuclear warheads into orbit, i.e., to strike any country on the globe, using one of its launchers.

In response, the US did not sit idly by.

They immediately accelerated their own space program by building the Vanguard TV3 satellite. As we know, due to a series of delays, the mission failed: Vanguard was lost in the very first launch phase. Only on January 31, 1958, would the United States succeed in sending into orbit the first satellite, Explorer 1, followed, on March 17 of the same year, by Vanguard 1.

But the Soviet successes aside, not everyone's consciences were unaffected by the death of Laika. In every part of the world, there were protests at the Soviet embassies. A large segment of public opinion seemed to be against the use of animals for scientific purposes.

But the ethical debate did not ultimately matter to the Soviets, even at the cost of being unpopular.

It was most important to be the first into space, with an animal, with a man, with a woman. Just first. In other words, before the US.

As to the fate of Laika, only at the end of the Cold War were we able to learn more. In 2002, the results of new research carried out by the Russian scientist Dimitri Malashenkov were made known: he reported that Laika survived only between five and seven hours after take-off, due to severe fluctuations in the temperature from hot to cold. But newer versions paint a still different picture: in this telling, it is said that her death was due to asphyxiation, resulting from a failure of the ventilation system.

The person in charge of the mission, Oleg Gazenko, in an interview in 1998, stated that he regretted the death of the animal, as it was an almost useless sacrifice. According to him, the mission gave them little information, in part because of the premature death of Laika, whose body was incinerated five months later, during the re-entry of the satellite into Earth's atmosphere.

Laika, however, became one of the most famous animals in the world. She was remembered among the cosmonauts who died on missions and has been celebrated in various tributes in popular culture. Stamps, songs, movies and videos have been dedicated to her.

In 2002, with the release of the album *Memoryhouse*, the musician Max Richter dedicated the song "Laika's Journey" to the first living being in space, sacrificed unnecessarily in the US/USSR's competition.

Russian biologist Adilya Kotovskaya was the last person to make contact with Laika on the eve of her launch.

In 2017, at the age of 90, the woman recounted:

I asked her to forgive us, and I cried as I petted her for the last time.

The next day, on November 4, 1957, Laika became the first terrestrial being to orbit the Earth.

After only nine orbits around the Earth, in her little box, Laika died, either from overheating or asphyxiation, alone and insane with fear. It didn't go as planned: she didn't orbit the Earth for eight days before the last portion of food—the one prepared with poison—was to put her to sleep without suffering. At first, Kotovskaya recounts,

> ...the mission went smoothly: at launch, Laika's heartbeats sped up considerably, but after three hours everything was back to normal.

So much discordance! What, ultimately, is the truth? We are still wondering.

But now, the writer would like the reader to answer a question before continuing.

Let us assume the existence of a superior being, much stronger than us, who can dominate us and who pretends to be our friend. This being forces us, without asking for our consent, to act as guinea pigs to test the survival of living beings (different or equal to him) in an unknown environment, moreover, without the possibility of returning alive to the self-proclaimed "friend." Would you forgive such a superior being? Or would you condemn him without the possibility of appeal?[9]

Here is one of the many versions. Audio first recorded the dog's yelps of fear. It was unimaginable, what was heard. After three hours, however, Laika seemed to calm down. She was able to eat, as she could hear the crackling of the gelatinous food that she had available.

But suddenly, during the tenth revolution around the Earth, the temperature inside the capsule began to rise, eventually exceeding forty degrees. Perhaps the capsule was not sufficiently insulated from sunlight. In a few hours, Laika found herself dehydrated. Soon after, the poor girl died.

But Soviet radio lied (this would not be the first or last time; we will see it happen with Valentina Tereskhova and on many other occasions), continuing to transmit updates every day. Everything was going very well. Everything was going on according to plan. As scheduled, the spacecraft finally crashed into the atmosphere five months later, on April 14, 1958. It crumbled over the skies of the Antilles. Laika, however, had not suffered—it was said—because she had been killed by an equivalent of lethal injection...

The official version was repeated for a long time: the (perhaps) true story of Laika came out only many years later.

---

[9] This question is very much in step with the current time. At that time (or even as late as 40 years later), these kinds of ethical considerations were not a factor within the scientific environment. They were just not there. Scientists conducted their experiments and mostly did not question whether it was ethical to use animals. Animal rights activists were a thing of the West; they did not start appearing in Russia until the '90s.

Laika was chosen because she was photogenic. She was needed for propaganda. She was also chosen because of her gender. Kotovskaya explains that:

> We chose a female because, to urinate, she didn't need to lift a leg, and this allowed us to save space.

Laika's feat only served to prove that a living being could survive the launch.

The Soviet Union had abolished the idea of a (distant) deity and replaced it with a much more approachable idea: the possibility of becoming HEROES. Even in the collective unconscious of the Soviet leaders, the Greek myth of the hero—son of a god and, thus, a demigod—i.e., a being endowed with special powers—was alive. In the Soviet Union, the concept of the national hero was systematized, because of its evocative power. Today, the Heroes of the Soviet Union, renamed"Heroes of Russia," number more than twelve thousand and wear a gold star. If the Russia of the Tsars had created its own nobility, with hereditary titles, Soviet Russia created Heroes.

Let's remember that the great anthropologist Vladimir Propp was Russian, and was author of *Morphology of the Fairy Tale*, a work that admirably describes how all fairy tales and stories have a HERO, an ADJUVANT, an OPPONENT, a MAGICAL MEANS and a REWARD.

Well, in the Russian space enterprise, the hero was the cosmonaut, the adjuvant the USSR Government, the opponent the United States of America, and the magical means the spacecraft, with the reward being to enjoy the privileges (financial and prestige) of a celebrity—for the immediate present—and to be remembered in this history books—for the future.

Winning this or that award? Being decorated with such and such medals? This is nothing compared to the possibility of becoming a **hero**: not a human being better than others, but a category in its own right. Who wouldn't want to be a hero?

After the death of Laika, the Soviet engineers had to prove that a living being could be brought back to Earth alive. In fact, other dogs were launched into space, and in August 1960, the little dogs Belka (Russian for "Squirrel") and Strelka (Russian for "Arrow") were sent into orbit aboard the Korabl-Sputnik spacecraft. The two dogs were nervous during the launch, but calmed down when the rocket reached orbit and stabilized. They breathed, ate normally and were successfully returned to Earth. They were the first living beings to return safely from a space mission. Then, it was the turn of a gray rabbit. Then, dozens of mice, flies, plants and fungi left Earth and all returned alive after a day in space.

It wasn't until fifty-six years later that the Cosmonaut Museum revealed the true identity of Belka and Strelka: they were actually called Kaplja (meaning "Drop") and Vilna. Two pet names, which Russian scientists decided to replace with their more evocative "battle" nicknames. According to the Russian newspaper Mk.ru, the discovery of the names of the two dogs was made thanks to the notes of Dr. Oleg Gazenko, who had reported all the biological parameters, both before and after sending them on an expedition around the Earth.

Kaplja (Belka) subsequently had a puppy, called Pushinka. In 1961, Belka's son was given by Nikita Khrushchev to Caroline Kennedy, the daughter of President John Fitzerald Kennedy.

But we do not want to end the history of Laika without mentioning an article that caused a stir at the time.

While, in Italy, some people were full of praise for the operation, imagining that Laika would have been happy to be the first dog to pee next to a star, the Italian writer and journalist Dino Buzzati[10] thought differently. In fact, in an open letter to Salvador de Madariaga, the Spanish diplomat, politician and writer, he wrote:

Illustrious Mr. De Madariaga,

We have read, with the pleasure that can be obtained from the ingenious and elegant joke of a great gentleman of European culture such as yourself, the article published last Tuesday by the 'Corriere della Sera,' in which you imagine a dialogue between the dog Laika locked in the satellite in flight and an obscure British dog... Laika does not find it cruel that men have thrown her into the sky with Sputnik, on the contrary, she is very pleased and feels that she has been honored with the part of a pioneer...

Well, illustrious De Madariaga, with all the consideration you deserve, I suspect that, this time, you have let yourself be carried away by literature. To imagine, as you do, that the tremendous task assigned to her would make Laika proud and cause her to exalt is synonymous with absurdity. Laika happy to explore space first? Laika intoxicated with speed? Laika satisfied with "making no effort to breathe"? Laika pleased with her perfect heartbeat?

But no one came, no hand caressed her throat, her moans were not heard by the perfect devices of the Soviet observers. Only God heard them, poor beast. So much for the greed of science!

On the other hand, at the very end of the article, a symptomatic oversight escaped you. Nothing serious, mind you. A zoological accident of tiny importance. There, where "your" Laika says, "Just think, a dog that for years has been content to raise

---

[10] Dino Buzzati, "Corriere dell'Informazione," November 16–17, 1957.

its hind leg against a gas lamp post can now do the same against a real star!" The image is brilliant and pathetic... But, sadly, it's wrong. Never, in her life, did the dog Laika raise a paw against lampposts, walls or lawns. Does it need to be added why?

And he concludes:

Farewell, then, gentle little dog, who no longer wags her tail, who will no longer have a kennel, I'm afraid, nor a lawn, nor a ball, nor a master. You will die in cruel solitude, without knowing that you are a hero of history, a symbol of progress, a pioneer of space. Once again, man has taken advantage of your innocence, he has abused you so as to feel even greater and to give himself a lot of airs.

Nonetheless, the feat of the dog convinced the Soviet authorities. Sending a human being into orbit was very risky, but it could be done. In fact, it had to be done. Once Laika had been liquidated, in April 1961, it was Yuri Gagarin's turn.

1.7   The Nedelin Catastrophe

The launch pad accident that occurred on October 24, 1960, at the Baikonur Cosmodrome, is known as the "Nedelin Catastrophe", and it happened during the development of the Soviet R-16 intercontinental missile. Beyond a few rumors, never confirmed or, for that matter, denied, the episode was kept hidden until at least 1989, and, although it is often associated with Soviet cosmonautics, it is fully a part of the military effort known as the "Cold War" that required the US and the USSR to have weapons so powerful as to be able to annihilate the opponent. It was necessary for both sides to make known to their rival that a certain weapon was available and that it would be used if necessary. At that time, the USSR did not have a ballistic missile capable of delivering nuclear warheads and dropping them on US territory. The R-16 was designed for this purpose, but, among the various versions, there was also talk of a launcher designed for a mission to Mars.

There is no one who has better studied, and in greater detail, what happened and how the "Nedelin Catastrophe" came to be than James Oberg. An American NASA engineer who later became a journalist and writer, Oberg is considered one of the major Western experts on the Soviet space program, which he was able to study, in part, thanks to his knowledge of Russian.

The following is the entire article that James Oberg published in *Air & Space Magazine* from December 1990. On the website www.jamesoberg.com, you can find countless other articles and studies conducted by Oberg in the most disparate fields of cosmonautics and astronautics, including the "lost cosmonauts" theory and the "lunar conspiracy" theory, as well as skeptical remarks about UFOs.

The sources used are quite disparate and, in some cases, conflicting. In order to better understand what happened, Oberg's article will be interrupted in the middle in service of a chronological reconstruction of the facts.

The bodies that could be identified numbered several dozen, including that of the officer whose poor judgement had caused the disaster. They were shipped home from the Soviet central Asia launch site for individual internment. Dozens more were burned beyond recognition in the horrible conflagration, and whatever remains could be found -- teeth, charred leather, shards of bone, keys and coins -- were swept up from the scorched concrete, placed in a single coffin, and lowered into a grave in a park in the rocket workers' city of Leninsk.

The families of these Soviet rocket workers were alone in their grief. Officials quickly announced that the commander had died in an airplane crash. As far as the rest of the world knew in that fall of 1960, the Soviets' efforts in space continued to move from one crowning success to another.

European journalists in Moscow soon picked up rumors that a gigantic rocket had exploded "in Siberia," killing hundreds, but those stories quickly took their place amid other oft-embellished legends of dead cosmonauts, super weapons, and similar folklore. U.S. intelligence officers had something more concrete: several blurred, spotty photographs of the site brought back by a Discoverer recoverable reconnaissance satellite. ("The scorched area was tremendous," one officer told me two decades later shaking his head.)

But at the time they were as quiet as the Soviets about their findings. Something horrible may indeed have happened, Western experts concluded, but there was no way to be sure what it was.

Time passed. The grave site in the Leninsk park was covered with a grassy mound 40 feet across and fenced in. Local officials erected a memorial obelisk, with 54 name-bearing plaques spaced along the four sides of its square perimeter. Friends, relatives, and co-workers at the Baikonur Cosmodrome launch complex kept the memorial decorated.

Other disasters occurred at the Cosmodrome from time to time, and new memorials were added to the park. One touching tribute was built in a corner of the spaceport's museum -- until recently kept secret from outsiders both Soviet and foreign -- where a scorched notebook found on an engineer's body was displayed behind glass. No label was necessary. Over the decades the local rocket workers, who knew the Cosmodrome's full history from first-hand accounts of survivors and family members, wore the wooden case smooth with their hands.

The recent opening up of the Cosmodrome to outsiders also opened up many of the workers' bitterness at the decades of official denial. "If you only knew of all the explosions and deaths," one museum official lamented to a visitor earlier this year, "you would be horrified at the size of the deceptions." Evidently much more is still held in secret Soviet archives or, worse, was documented in records the museum staff was regularly ordered to destroy. But none of those later accidents at the Cosmodrome (or another that killed 50 men at the Plesetsk rocket center north of Moscow in 1980) ever approached the death toll of that October evening only three years after Sputnik 1.

Over the years, many conflicting accounts of the disaster reached the West. As a lifelong space nut fascinated with Soviet mysteries and the sleuthing needed to unravel them, I collected and evaluated the stories and tried to fit the pieces together for more than a quarter of a century. Details came from credible Soviet sources both inside the USSR and overseas. Top-level spy Oleg Penkovskiy, executed in 1965, wrote in his memoirs that a "nuclear-powered" missile had exploded, and many recent Russians elaborated on the theme (apparently basing their reports on the coincidental deaths of several top Soviet nuclear weapons experts elsewhere that October). Zhores Medvedev, who had a record of correct assessments, reported that the disaster involved a "moon rocket" needed for a propaganda spectacular. Nikita Khrushchev himself mentioned the disaster in the first volume of his memoirs, smuggled out of the Soviet Union and published in the United States in 1970, but he gave no hint of the role he may have played.

From these stories a scenario emerged.

What were the steps through which it all came to pass? Here is how the accident has been reconstructed through data and information taken from various sources and assembled over the years, in particular, from the article "The Nedelin Incident."[11]

The rocket was defective. This was in the second half of the '50s, and in the space race, the Russians were achieving one success after another, one record after another, in the face of the US and the entire West. The Russian space program has always had a wild card: the very safe R-7 rocket, a powerful symbol of Soviet astronautics, compact, 100% safe, wonderful for in-orbit missions and for delivering scientific material during interplanetary journeys. But the Russian space program was only one sector of the Soviet missile program. After all, the military was indifferent to getting things into orbit and returning them to base in perfect condition. In fact, they only cared about the "things" if they were nuclear warheads, and better yet if they were aimed at a US population center.

---

[11] https://leganerd.com/2014/12/11/lincidente-di-nedelin/.

Many experiments were carried out to see if it was possible, on the civilian side, to bring an intact capsule back to earth and, on the military side, to reach a target to be destroyed without the cargo burning up first. All of this was feasible only in 20% of cases, a low percentage at the civil level and disastrous if you wanted to have the upper hand in a nuclear war.

As if that were not enough, there was also the unavoidable constraint of time. The very reliable R7 had to be charged with kerosene/liquid oxygen: it took twelve hours to do this. Moreover, if, for any reason, the launch had to be cancelled, the rocket had to be unloaded and then reloaded.

All of these things were not serious in and of themselves if it was a matter of sending someone into orbit. But they became so in the case of a nuclear counterattack. To add to the problems, there were the launch bases. The only places from which such missiles could be launched were the Baikonur and Plesetsk cosmodromes. But, from a military point of view, they were indefensible: in the case of a US attack, the Americans would comfortably be able to demolish the launch pads and the Soviets would be left in the lurch, with no possibility of defense.

The Strategic Missile Forces needed a reliable missile, that is, one that could be kept hidden and safe for months in an armoured silo, with full tanks, in a state of alert, and, above all, without the need for maintenance: ready even to be loaded and launched in ten minutes from the arrival of orders. Not bad at all! The army had partially avoided any difficulty with the use of nuclear submarines and short range missiles; moreover, it had developed the R-12 and R-14, which were capable of at least reaching Europe. But the US had Titan-1 missiles ready, and had buried them in various areas. It was enough that they only had to be aimed at the Russians, once the Russians had given them an excuse to attack. At that point, it was necessary to develop a missile that could address these significant difficulties.

In May 1958, Khrushchev wanted a new carrier rocket to be invented. An elegant two-stage missile, OKB-586 design.

Khrushchev had asked Korolev for it, but the latter had said no. Korolev was a very talented man, but his interest was cosmonautics and related research, so he was perfectly willing to use military funds to design missiles, but only if they were related to space research.

There was, however, another name to be considered: Mitrofan Nedelin, who had long wanted to overtake Korolev. Nedelin agreed to Khrushchev's request and began working with the design team to come up with a perfect ballistic missile, i.e., one that could meet military indications and requirements. And he succeeded. The working collective of designer Mikhail Yangel eventually arrived at the design of the R-16 missile.

And it was thanks to Marshal Mitrofan Nedelin that the military appeared in that episode. It was he who had persuaded Khrushchev that the war resources of the future would be based on missiles. In that view, many military and naval factories had been converted for the production of missiles, meaning that, at that moment, the army found itself debilitated and with no possibility of rebalancing. Khrushchev was furious when he found out. The only way for Nedelin to come out of this with honor was to develop a suitable intercontinental missile soon.

Thus began the R-16 project. The missile was more than 30 m long, three meters in diameter, and had a launch weight of 141 tons. But the fuel, as some warned at the time, was a recipe for disaster.

The rocket was equipped with two stages and used new engines that burned hypoazotide and asymmetric dimethyl hydrazine, two highly toxic and volatile compounds, which, however, had the undoubted quality of being liquid at room temperature (i.e., easily storable, since the high boiling temperatures ensured the preservation of the fuel and the comburent) and hypergolic, i.e., ignited by simple contact, without the need for a complex ignition system. It was enough to blow up the valves of the tanks (they were pyrotechnic and, once opened, could not be closed) and the rocket would start. An easy and guaranteed procedure, a dream of all army generals. The rocket was a little less powerful than the R-7, but it was able to stay underground for a long time without causing difficulties, to be equipped within minutes and, above all, to launch a 3-tonne atomic device at the United States! That was all the military wanted. It must also be said that these were toxic and carcinogenic fuels, so much so that the supply personnel were forced to wear hazmat suits.

In the meantime, Khrushchev continued to put tremendous pressure on Nedelin.

So, the R-16 project went ahead, progressing with incredible speed. So much so that, in October 1960, the first prototype was ready to be tested.

The project proceeded in forced stages and many safety protocols were skipped. In a context of substantial emerging engineering difficulties, prelaunch testing even began to overlap with preparations for launch.

Everyone was so nervous! While Nikita Khrushchev anxiously awaited the arrival of another playing card in the Cold War with the United States, Mikhail Yangel, the chief designer of the R-16 developed by his collective, couldn't wait to show Khrushchev what he had achieved.

Later, surviving veterans would unanimously testify that the rocket had been plagued by problems since the day its tests began at Tyuratam—another name for Baikonur—on Sept. 26, 1960.

It is true that it was nothing new for the Soviet rocket industry to have to speed up, but, until then, things had been done quickly, but carefully. But in this case, too many interests were at stake and the work was carried out in an environment where the tension was palpable. Many of those who worked on it said, after the disaster, that they knew of no other projects carried out with the same fury, the same eagerness to skip steps, as long as the objective of reaching launch in the shortest possible time was achieved.

The prototype was brought to the Tyuratam ramp on October 21, 1960. Once the rocket had been moved, the guidance system was not even tested, and many of the electronic systems were totally untested. All of the inconveniences that cropped up in the initial tests were underestimated. And so, the command to go ahead was given without any preceding investigation into the reasons for said inconveniences.

Thus, on the evening of October 23, 1960, the rocket was on launch pad 41, ready to go. The rocket was surrounded by dozens of technicians rushing to get the last details ready before launch. Then, the order came, and the pyrotechnic tank valves began to pop one after the other, connecting the compartments of the two fuels.

But the rocket did not ignite.

After the first few moments of apprehension, they began to try to understand what had happened. The valves had been blown. The rocket was on the ramp, but the engines had not ignited. The hypothesis was put forward that one or more of the valves had not triggered, so the reaction had not been initiated. Which valves had misfired? Impossible to know; the status of the valves was not among the data communicated via telemetry. So, there were no automatic controls to figure it out. The technicians were not sure of the number of explosions actually heard.

The State Commission subsequently arrived to decide what to do.

The explosive cartridges needed to be changed, and probably the first stage control unit as well (that, too, could have had some drawbacks). But some of the valves were already open. Their components were highly corrosive, so they were wearing out the seals. The seals could only work for twenty-four hours at the most.

So, it was necessary to hurry. And there was another difficulty: regulations for emptying the tanks meant it would take that much longer to do so. It was a very delicate job, because a small leak could cause a disaster.

The Commission came to this conclusion: the only really safe thing to do was to blow up the isolation valves of the tanks, so as to make them safe, and then deactivate the first and second stage control units to prevent them from sending faulty commands, after which they would empty the tanks, clean the pipes, change the faulty parts and re-run the test.

That way, there would be no risk.

These were the common sense things that should have been done. But that would have taken days. Too long for Nedelin, who was under pressure from the constant phone calls from Khrushchev and many Kremlin dignitaries. So, Nedelin, supported by the commission that supervised the test, decided not to allow for any delay. Experienced engineers and military personnel chose to ignore the most basic safety rules: repairs would be carried out on site, i.e., on the rocket full of fuel, on the launch pad.

The danger to personnel was frightening.

Nedelin even had a desk placed twenty meters from the ramp for himself. The man took his place with the other commissioners and began examining the documentation. Seeing the boss set an example, the technicians did not dare protest, and returned to their work in a disciplined manner.

Ultimately, it is not surprising that he would choose to take such risks. Nedelin was a decorated artillery officer, a Hero of the Soviet Union, one of the great supporters of the development of the missile weapon… His career was at stake with that launch!

Nedelin even went so far as to demand that the faults be fixed while the missile was in launch position, so that the tests could be carried out the next day.

There was a big fight over this, but Nedelin remained adamant. He demanded that his pleas to go ahead be heeded and gave orders to begin repairs. The repairs went on all night, by the light of the photoelectric cells. Soldiers were present to assure that the work continued and the poor technicians took care of the breakdowns. And all the while, everyone stood around that great metal beast, loaded with hypoazotide and dimethyl hydrazine and ready to be launched.

Neither the head of the NIIP-5 test range at Tyuratam, Major General Konstantin Gerchik, nor the head of the 2nd Test Directorate, Grigoryants, who was in charge of the R-16 tests, could enforce the safety rules in the presence of Nedelin, their manager.

Among other problems that arose, the launch staff had discovered a fuel leak aboard the rocket, estimated at 142–145 drops per minute. However, technical management had announced that the leak was of no concern, as long as it was contained. Launch managers had assigned a chemical unit staff to keep an eye on the leak. Boris Konoplev, the chief designer of the control system, personally supervised the checks from a position inside a bus parked on the launch pad.

As launch time approached, the State Commission members gathered at the IP-1B ground control station at Site 43. At 8 a.m. local time, Oct. 21, the LD1-3 T vehicle left the assembly building at Site 42 and, an hour later, the rocket was installed on the "left" launch pad at Site 41.

A wooden deck was prepared there so that the committee could view the launch.

However, when another 30-min delay was announced, Nedelin, apparently under pressure from Moscow, asked to be escorted to the launch pad "*to see what is going on.*" Eyewitnesses say that, at least twice on October 24, Nedelin received calls through special communications channels from the Kremlin, possibly from Khrushchev himself. The calls always wanted to know when the R-16 would fly. An estimated 250 unsuspecting people were still around the rocket at 6:45 p.m. local time, about 30 min before the scheduled launch. There are still conflicting accounts of Nedelin's behavior on launch day. His closest associates defend his decision to go to the pad as an example of his dedication to the program. However, other veterans of the program say Nedelin's attitude only distracted staff and further compromised safety at the pad.

In addition, the membranes on the fuel and oxidizer lines of the second stage had been activated, so the components of the self-flammable propellant were very close to the engine combustion chamber.

On the morning of October 24, however, it was suspected that the second stage control unit was also faulty, so it was decided that they would blow up all the safety valves in advance: in this way, as soon as the first stage was released, the control unit would turn the second stage on with a single command, regardless of anything else. It was a tremendous risk, and there were many other flaws as well. But most of all, people were exhausted and nobody remembered a very important detail: the second stage control unit was still on. True, it was without power, but its internal batteries were enough to make it work.

Half an hour before the launch, the various systems were restarted from scratch, including the surveillance cameras of the polygon. At that point, the second stage control unit, which had remained active, was also powered up, and the countdown began. It started at 90s, just as the technicians were checking the last parameters.

90s later, the clock reached 0 and the control unit started the second stage. But the rocket was still sitting firmly on the ramp when it should have been miles from the ground!

In fact, since there had been no additional control and the safety valves had already opened, the control activated the automatic systems that ignited the engines of the infamous second stage.

The metal underneath melted and the fuel in the first stage tanks ignited, setting off the greatest disaster in the history of space exploration, and of rocketry in particular.

A blaze of indescribable size hit the first stage, and the metal partitions resisted for a few seconds before being destroyed: at that point, tens of tons of lethal and explosive fuel spread across the ramp.

The fuel immediately ignited, coming into contact with the flames of the second stage. A tremendous 3000-degree blaze rose up, a fireball that expanded like a wall, charring everything within a hundred meters and burning alive the dozens of people who were still standing around the rocket. Others were killed by poisoning and chemical burns caused by the fuel vapors.

Firefighters at the range battled the blaze for two hours before they were able to get through. Only then were rescue workers able to intervene. But there was little left to rescue.

However, there are other versions of this story. In one, the ignition of the second stage engine during the pre-launch test was caused by a short circuit in the replaced main sequencer. This caused the first stage fuel tanks to explode directly below, destroying the missile with a huge explosion. Before seeking shelter, the camera operator remotely activated several automatic cameras positioned around the launch pad, which filmed the explosion in detail.

People near the rocket were immediately incinerated; those further away were burnt to death or poisoned by the toxic fuel vapors. Andrei Sakharov described this picture: as soon as the engine started, most of the personnel rushed to the perimeter, but were trapped inside the safety fence and then engulfed by the fireball of burning fuel.

Meanwhile, in the control bunker, General Matrenin yelled to his subordinates not to touch the main control panel, trying to preserve the position of the switches at the time of the incident. At that point, a burn victim burst into the bunker, and the others rushed to his aid. After an unsuccessful attempt to close the bunker door, Matrenin ordered the personnel to put on gas masks. When the explosions subsided, he ordered the survivors out of the bunker and to the launch complex checkpoint.

The explosion had now incinerated Nedelin, along with one of his best aides, the USSR's top missile guidance designer and more than 70 officers and engineers. Others died later from burns or poisoning.

By a stroke of luck, the missile designer Mikhail Yangel and the commanding officer of the test range survived, suffering only a few burns, as they had gone to smoke a cigarette behind a bunker a few hundred meters away. They had been invited to take a break by General Mrykin, along with Iosifiyan, a leading electrical systems engineer, who also convinced Bogomolov, a non-smoking colleague, to go with them, presumably to discuss the situation.

According to the memoirs of Boris Chertok (Russian engineer, designer of Soviet control systems, and author of the four-volume book *Rockets and People*—the definitive source of information on the history of the Soviet space program), Iosifiyan and Bogomolov hoped to convince Yangel to delay the launch once again, and to stop conducting repairs on a rocket fully loaded with fuel.

After the explosion, their colleagues had to restrain Yangel, who tried to throw himself back onto the pad, possibly in a state of nervous shock.

The exact death toll from the explosion is not known. In 1965, the spy Oleg Penkovsky passed to the West the information that the dead numbered 300. The Soviet Union said only that a "significant number" of people had died, acknowledging the incident for the first time in April 1989. Later in the year, the government set the death toll at 54. The most recent estimated death toll, published by Roscosmos on the 50th anniversary of the accident and researched with engineer Boris Chertok, was that 126 people had died, but the agency concluded that the actual number could be between 60 and 150 deaths.

Ultimately, the committee discovered that there were many more people on the launch pad than there should have been: most should have been told to return safely to the bunkers.

Dozens of technicians, workers and soldiers met a terrible end, consumed by fire. But how many exactly? Perhaps no one will ever know. The official reports were declassified in 1994, but the total number of victims was never precisely established, since many were literally dissolved and vaporized by the terrible explosion. However, people working at classified jobs had to sign in and out of the facility each day, so it is certain that the exact number and names of people must have been known, due to that established and strictly followed procedure.

The investigation commission, headed by Leonid Brezhnev, then 'ceremonial president' of the USSR, established that, during the manufacture of the prototype and during pre-launch operations, an incredible amount of violations of safety protocols had accumulated. So many procedures had been circumvented, so many circuits disconnected.

When buses sent from Site 10 returned with the first survivors, officials refused to tell the local hospital what kind of "secret" chemical had poisoned the patients.

However, the commission determined that those responsible should not be punished: they had already paid for most of it with their lives. When Brezhnev arrived at the firing range on October 25, 1960, he said:

> Comrades! We do not intend to prosecute anyone; we will investigate the causes and take steps to recover from the disaster and continue operations.

Despite this, according to some sources, I. A. Doroshenko was held responsible for the event.

Later, when Nikita Khrushchev asked the designer Yangel, "*But how did you stay alive?,*" Yangel replied, in a trembling voice, "*I went to smoke a cigarette. It's all my fault.*"

Among the wounded who survived the fire was General Gerchik, who was delivered to the hospital with extensive 2nd and 3rd degree burns. According to his own recollections, he was right next to the missile at the time of the explosion, checking the situation with the leak. Gerchik claimed that a gust of wind had driven the flames away from him. In total, 49 survivors were delivered to the hospital and placed in intensive care. 16 of them would die within several months.

However, the government immediately made sure to avoid any leakage of information.

The bodies of the soldiers were all buried in a mass grave not far from the cosmodrome, while the bodies of the 17 highest ranking soldiers were returned: 17 accidents were staged—all on the same day or close to it—to make the cause of their disappearance appear to be as far away from the truth as possible. All survivors were ordered to keep quiet. Soon, however, rumors of a gigantic catastrophe began to spread, thanks to the information gathered by Western intelligence services.

However, the news remained very imprecise, both on the size of the incident and its nature. Some uncertainties were dissolved only when the official documents were finally declassified.

Let's continue with James Oberg's story:

> *Late one afternoon a rocket's countdown was halted when problems cropped up. The launch team, ordered outside to attempt repairs, mounted the scaffolding around the balky, fully fueled missile. Suddenly the second-stage engine ignited, bursting the fuel tanks of the first stage and covering the launch pad in a tidal wave of flame.*

In my books, articles, and lectures, I labeled the event "the Nedelin Catastrophe."
If any one man deserved such an eponymous disaster, it was Field Marshal Mitrofan
Nedelin, then 54 years old. The commanding officer had violated all standards of
safety when he ordered the technicians onto the launch pad. Perhaps to support
his order he went outside himself, and he died with the others when the missile
exploded.

I tried to add it all up. I knew that two unmanned Mars probes had been
unsuccessfully launched only two weeks before from the pad used to launch Sputnik,
and I believed that the basic Sputnik booster, the R-7, was the only big Soviet
rocket flying at the time, so I postulated that the rocket that had blown up was
also a Mars-bound vehicle. The Soviets generally prepare three vehicles for any
major space effort. The pressure on Nedelin to launch would have been intense:
Khrushchev had been at the United Nations in New York earlier that month
giving a speech about Soviet foreign policy and anticipating another spectacular
feat to flaunt before the world. Furthermore, the launch window—the planetary
alignment that allowed such launchings—would have been rapidly closing day by
day. That was the scenario I proposed in my book about the Soviet space program,
Red Star in Orbit, in 1981.

The Mars rocket scenario couldn't account for a few troubling items, however.
The ships used to track the Mars probes had been in position in the south Atlantic
and northeast Pacific for the failed October 10th and 14th launches, but they
had set course for their home ports before the explosion. There were also reports of
the involvement of rocket designer Mikhail Yangel, who was not a member of the
team behind Sputnik and the two previous Mars shots. Furthermore, the Sputnik
pad was used in a launch only five weeks after the explosion, suggesting little if any
damage there. Most tantalizing was the spy Penkovskiy's explicit reference to funerals
at a rocket plant in the Ukraine, an installation later revealed to be devoted entirely
to military projects.

By the time I updated the account for a new book, Uncovering Soviet Disas-
ters, in 1988, my belief in the Mars hypothesis was fading. As the book went to
press, I began to regret not offering a second hypothesis: that the rocket was an
intercontinental ballistic missile.

And still I despaired of ever finding out what really happened, short of the
violent overthrow of the Soviet government and a personal search through captured
top-secret archives. These pages of space history, I thought, were fated to remain
blank forever.

But my pessimism was overtaken by recent events in the Soviet Union. In 1989
the first published account of the disaster appeared. A magazine article by Aleksandr
Bolotin, a young officer at the Cosmodrome, in the pro-glasnost weekly Ogonyok,
identified the rocket as an ICBM. More than confirming my suspicions, the article
personalized the horror for me. When it mentioned a memorial obelisk over the
burial site, I promised myself that someday, somehow, I would visit it.

Early 1990 found me before the obelisk, reading aloud the names of the dead
and placing a bouquet by the stone. Standing there in the mid-winter gloom,

brushing the snow off a few of the plaques, I did not feel like the winner in some "pierce the cover-up" contest. Rather, I was pleased that after 30 years, a rip in the fabric of reality was finally being repaired. A feeling of wholeness, of a fully restored flow in a history long obstructed, made me proud to have played a small outsider's role in the mending process.

I had arrived at the summit of my investigation thanks to a project on the Soviet space program for PBS' NOVA television series (to air in the United States this February), for which I served as researcher and on-camera tour guide. Getting to the Cosmodrome last February was difficult, but the crew and I surmounted the bureaucratic obstacles and at last arrived at Baikonur. We asked for a van to take us to the memorial park I'd spotted from the bus on the way in from the airport, and we had to spend half our free afternoon pleading and prodding for it. Persistence finally paid off.

Many of the plaques were cracked with age, but the shrine had not been ignored. Flowers, pine boughs, and tufts of prairie grass decorated many of the markers. I asked my guide who had made these visits so many years after the explosion and his reply caught me by surprise: "Weddings." Since the rocket workers' city has no World War II memorials like the ones newlyweds traditionally visit and decorate elsewhere in the U.S.S.R., the obelisk had assumed the role. Several times a week, groups of young people came on foot from wedding ceremonies in Leninsk to stand by the grave site, pause in thought, and honor their dead.

But every answer raises another question. I might have taken the 54 names listed there as the total death toll had I not noticed that Nedelin's name was not among them. When I asked my guide why, he replied that the commander's body had been sent home for burial. How many others had been sent home? I asked. The guide thought for a moment "About 40," he suggested tentatively. The death toll, then, was nearly 100 men.

The details of the disaster were confirmed and elaborated on by the Ogonyok article, the only one ever to appear. The designer Yangel was in charge of the technical proceedings at the pad, it said, but at one point he became so nervous he stepped into a special fireproof hut for a cigarette. It was while he was inside that the rocket exploded, probably when a technician plugged the first stage's umbilical cable into the second stage's receptacle, causing a normally innocent command wire to trigger the ignition.

Yangel survived by a fluke. But many of the USSR's spaceflight pioneers perished in the accident. One man named Nosov had pushed the launch button for Sputnik three years earlier; another named Ostashev had been instrumental in developing the Sputnik booster. In Leninsk there are streets named "Nosov" and "Ostashev" among the usual "Marx," "October," and "Red Army" streets.

As I stood before the obelisk, the gruesome details in the Ogonyok article came to mind. The explosion had occurred on Monday, October 24, shortly after 6:45 in the evening. One man who miraculously survived gave this account: "At the moment of the explosion I was about 30 meters from the base of the rocket. A thick stream of fire unexpectedly burst forth, covering everyone around. Part of

the military contingent and testers instinctively tried to flee from the danger zone,
people ran to the side of the other pad, toward the bunker ... but on this route
was a strip of new-laid tar, which immediately melted. Many got stuck in the hot
sticky mass and became victims of the fire .... The most terrible fate befell those
located on the upper levels of the gantry: the people were wrapped in fire and burst
into flame like candles blazing in mid-air. The temperature at the center of the
fire was about 3,000 degrees. Those who had run away tried while moving to tear
off their burning clothing, their coats and overalls. Alas, many did not succeed in
doing this."

Another witness had been on the pad but had finished his work and been ordered
away by Nedelin. He went to the observation point on a small hill about two
miles away, where a crowd of officers and engineers was relaxing. "Above the pad
erupted a column of fire," he recalled. "In a daze we watched the flames burst forth
again and again until all was silent." He rushed to the medical center to help the
survivors and found the front of the building surrounded by bodies. "All the bodies
were in unique poses, all were without clothes or hair. It was impossible to recognize
anybody. Under the light of the moon they seemed the color of ivory." It was a long
time ago and the bodies had been at rest for decades, but standing at the obelisk I
felt a chill down my spine.

Andrey Sakharov's newly published memoirs add a poignant detail to the tragedy.
When the accident occurred, he wrote, "automatic cameras had been triggered along
with the engines, and they recorded the scene. The men on the scaffolding dashed
about in the fire and smoke; many jumped off and vanished into the flames. One
man momentarily escaped from the fire but got tangled up in the barbed wire
surrounding the launch pad. The next moment he too was engulfed in flames."

What appeared to be authentic footage of the explosion aired on Soviet television
last April 12, "Cosmonaut Day." The films showed a rocket exploding and human
figures on fire running and falling. But the horror was not specifically identified or
connected with Nedelin, who is still, officially, a hero.

Back at the launch site, Nedelin's memory is not so dear. When I had first asked
to see "the Nedelin memorial" I was gently rebuked by my young guide, who hadn't
even been born when the tragedy occurred. "The monument is for all who died
that day," he said. There is no Nedelin Street in Leninsk either.

With these pages of rocket history blank no longer, I mused about the implica-
tions of the tragedy. The revelation that the exploding rocket was a military ICBM
puts the disaster into the greater perspective of the Cold War. Indeed, it can be
argued that the catastrophe almost led to a thermonuclear war. The Sputnik's R-
7 had turned out to be a great booster but a poor weapon. Only four were ever
deployed as missiles, at the Plesetsk military center. A second Soviet rocket team
was pushing hard for a new rocket to counter the Atlas missiles the U.S. was then
deploying. It was that missile, the R-16, that exploded.

Flight tests the following year were unsuccessful, probably due to the loss of so
many experienced engineers. By early 1962, as Americans began deploying ICBMs
in entire squadrons, Khrushchev was faced with a tremendous missile gap. It was

at this point that he decided to place missiles in Cuba, a gamble that brought the world to the brink of war during the Cuban missile crisis. But as space historian Curtis Peebles recently observed, the strategy would not have been necessary had the Soviets' new missile succeeded sooner.

Those speculations tugged at me as I walked around the memorial square and read each plaque's name. They were all Russian or Ukrainian, and most belonged to 20- and 21-year-old soldiers. I couldn't help thinking that their loss might have been more meaningful had it been for space exploration, the common world struggle that has claimed so many other lives around the planet. But these young men had died building a weapon, not a space probe.

I stood by the cold, lonely graves and tried to imagine the rocket workers' perspective, influenced by wars both hot and cold. They surely thought of themselves as defenders of their nation and as explorers too, since multipurpose missiles, such as the R-7, were being diverted to peaceful space activities. For no fault of their own they met a horrible fate. It had been my happier fate to spend three decades wrestling their reality from the denial and distortion wrapped around it. Now their nation was safe, and so was the truth.

In conclusion, what happened at the Baikonur Cosmodrome was kept secret until 1989, when the tragedy was made public. During the Cold War period, however, some news was leaked, reaching the United States: on several occasions, the continuous missions of U2 spy planes in Soviet territory photographed the Baikonur launch site. In the first photos, several rockets were seen on the launch pads. Later, there were photographic surveys that showed only large black spots on the ground, a clear sign of the scorched earth resulting from a serious explosion.

"The direct cause of the accident," reads the commission's final report, "was deficiencies in the design of the control system, which allowed the unscheduled operation of the EPK V-08 valve that controls the ignition of the second stage main engine at the pre-launch time. Problem not encountered in all previous tests. The LD1-3T vehicle fire could have been avoided if reconfiguration of the power distributor to the zero position had been performed prior to activation of the on-board power supply."

So, the summary is that the disaster happened precisely because of the rush and the pressure to launch the rocket fast. Because of that, the repair stage (fixing all of the bugs detected through testing, which were plenty) and the launch preparation stage were done together on a rocket fully loaded with fuel and in a rush that was strictly prohibited by the safety protocols. The active external power supply that was used on board for faster launch preparation, and that had been left there forgotten, allowed for the ignition of the 2nd stage engine during the routine of 'zeroing' of the control program. In order to set the program to zero, it had to cycle through all stages, including the start. When performed on unpowered circuits, it is inconsequential, but if the circuit is powered, it starts the launch, which is irreversible.

Safety concerns had been raised before the tragedy happened, precisely regarding unloading the fuel and carrying out all repairs and retesting before refueling the rocket. However, it turned out that **there was no technology for safely unloading the fuel.**[12]

After the accident, the R-16 program resumed in January 1961, and the first successful flight took place on February 2, 1961.

The Italian press first reported, on December 8, 1960, from undisclosed sources, that Marshal Nedelin and 100 people had been killed in a rocket explosion. On October 16, 1965, Oleg Penkovsky, the captured spy, had confirmed the details of the rocket incident, and exiled scientist Zhores Medvedev provided further details in 1976, in a British weekly newspaper. However, it was not until April 16, 1989, that the Soviet Union acknowledged the events, with a report entitled "Sorok Pervaya Ploshadka" (Russian for "Site 41"), which appeared in the weekly *Ogoniok*, the mouthpiece of PM Mikhail Gorbachev's "*perestroika*."

Every year, on October 24, elementary school students from the nearby city of Leninsk travel to Baikonur to lay flowers and fir branches near the memorial erected in memory of the Nedelin tragedy.[13]

As much as we may marvel today at the series of foolish decisions that led to the accident, we must remember that the era of rocketry, which today allows us such incredible successes as the Rosetta mission, was born and developed as a way to destroy opposing nations.[14]

The missile on the ramp was designed to incinerate a city of millions on US soil and to protect itself (through fear of reprisals) from a similar eventuality on Soviet soil. So, it was normal to act in great secrecy and sometimes sacrifice safety standards: a weapon of mass destruction was being tested.

The Nedelin catastrophe, which took the West by surprise, contributed to the legend of cosmonauts lost in space, launched by the Soviet Union on orbital and suborbital missions and never returned to Earth, due to accidents and malfunctions of their spacecraft.

---

[12] Here, we can find at least a part of the unclassified official report from 1960: https://www.livein ternet.ru/users/adpilot/post375063454.

[13] Here, we can find some original footage from the launchpad cameras, which, by the way, were turned on automatically once the launch program had started erroneously: https://www.youtube.com/watch?v=IHD2X-CvY6g.

[14] Here is another, pretty complete description from a historian of rocket technologies: https://war spot.ru/18227-nedelinskiy-koshmar.

Like every accident in rocketry history, this one had a lesson for the future. Today, when a rocket is on the launch pad and filled with fuel, a safety cordon is erected around it. This cordon cannot be crossed by anyone, except for the most urgent operations. This is a measure that has certainly saved many lives. The Nedelin disaster led, in spite of its tragedy, to many improvements along the road that brought humanity into space: the rockets we use today are the children of those designed to destroy our cities. And it is not the Russians or the Americans or the Europeans who are going into space, it is humanity as a whole doing so. Thus, those who died in the space race died for everyone. The deaths of the Nedelin catastrophe should be remembered for that as well.

1.8   The Tragedy of the Plesetsk Cosmodrome

The Plesetsk Cosmodrome is the northernmost Russian spaceport: it is located in the Archangel Oblast, about 800 km north of Moscow and just south of Archangel. In the beginning, the USSR had built it as a launch site for intercontinental missiles, with an urban area equipped with a railway station. Plesetsk holds the record for the number of rockets launched into space: more than 1,500 missiles have departed from its cosmodrome, more than any other launch site. But after the breakup of the USSR, the base was used much less. Its existence was a secret for a long time, but it was discovered by chance, thanks to the English physics teacher Geoffrey Perry and his students. In 1966, they studied the orbit of the Cosmos 112 satellite in detail and concluded that it could not have been launched from Baikonur. After the end of the Cold War, it became known that the CIA had suspected the existence of a launch base for ICBMs in Plesetsk since the late fifties. But it was not until 1983 that the USSR officially admitted its existence. This launch base is especially used for military satellites with high inclination and polar orbits, since the trajectory of any falling fragments would certainly be to the north, that is, a scarcely inhabited arctic and polar belt. But performing launches from Plesetsk involves a very high cost, unlike cosmodromes closer to the equator, which can launch heavier payloads. In this very cosmodrome, a tremendous explosion occurred that killed dozens of people. After the Nedelin catastrophe, this was one of the most terrible disasters in the history of cosmonautics. On March 18, 1980, there was a plan to launch the Vostok-2M rocket, carrying a military spy satellite. The rocket had a very good reputation: that type of rocket, in the previous sixteen years, had had only one accident. And after 1970, there had been no more problems.

Before the launch, all tests were run and no defect was found. The explosion occurred 2 h and 15 min before take-off. There were 141 people in the immediate vicinity of the rocket while it was being refueled. Tons of fuel caught fire. Fortunately, the launch pad workers were able to move the fuel trucks from the area, otherwise the tragedy would have been even greater. According to official sources, 44 people died as a result of the fire. Four others, seriously injured, were added shortly afterwards: they had been taken to the hospital, complaining of burns of varying severity. The State Commission, which investigated at the scene of the tragedy, found the victims guilty, but no criminal proceedings were opened. Fueling crews were blamed, but another independent commission, 16 years later, absolved them of any responsibility: the real cause of the fire was the use of catalytically active materials in the manufacture of hydrogen peroxide filters.

1.9   The Lost Cosmonauts

Luca Boschini is the author of the book *The Mystery of the Lost Cosmonauts—Legends, Lies and Secrets of Soviet Cosmonautics* published in the series of CICAP[15] notebooks in October 2013.

In that volume, the author talks about the legend of the "lost cosmonauts," a conspiracy theory, according to which Yuri Gagarin was not the first man to fly in space, but rather that there were missions before him that ended tragically with the death or serious injury of the pilot. The alleged failures were kept hidden by the Soviet authorities to avoid repercussions for their prestige in a very delicate period. The events we are talking about date back to the pioneering era of human space exploration, at the beginning of the '60s.

Here is what Luca Boschini had to say:

I work in Milan for an aerospace company and I am a member of CICAP. Moreover, I study Russian, so I could also read the original sources in their original language. The theory of the lost cosmonauts stated (through sources from the governments of countries that were in the communist orbit, like Czechoslovakia and Bulgaria) that the Soviet Union had already launched cosmonauts into orbit. But since they were not talking about it, it was assumed that something had gone wrong.

---

[15] From Wikipedia: Cicap is a non-profit educational organization founded in 1989 to promote scientific and critical inquiry into pseudoscience, the paranormal, mysteries, and the unusual, with the goal of disseminating scientific thinking and critical thinking.

He went on to say that another piece of the so-called proof was related to radio intercepts by Western amateur radio operators. Among the most famous were those of two people from Turin, the Judica Cordiglia brothers, who, with their equipment, intercepted radio communications from beyond the curtain and could provide journalists with first-hand news. The Judica Cordiglia brothers also had a famous space radio listening station in Turin.

There are a number of concomitant causes that have given rise to the theory of the lost cosmonauts. The first one is related to the fact that, for a while, the Soviet Union released almost no information; suddenly, after months of silence, they came out with a flurry of news: "First man in space," "First space walk," "First flight of two astronauts." So, the media had nothing to go on from sources closest to the story. The second one, related directly to the first, is the fact that American journalists would then subsequently look for information from the intelligence services of their own governments, which had every interest in indicating that the Russians were playing dirty by allowing cosmonauts to die in orbit. Another reason was radio intercepts, which were not always reliably interpreted. Other sources of information were refugees from the Soviet Union, people who had fled across the border and could tell what was going on on the other side, although these were often people reporting hearsay or third-hand news. Then, there were recordings thought to have come from ground stations. Boschini tried to correlate the dates of known interceptions with satellite launches (meteorological, communication or spy satellites). In fact, there are a couple of recordings that could coincide with the launches of American spy satellites. But the sounds recorded are quite strange, so strange as to seem to be wheezing or gasping, and it is unknown why such a satellite would have produced sounds like that. In many other cases, the communications in question turned out to have been received from airplanes on the ground, and were misinterpreted as having come from space. At that time, most of the communications that could be picked up by amateur equipment were in the HF band at a rather low frequency, below 30 MHz, where the atmosphere reflects the radio waves very well: this was a great advantage, because you could hear very far, as the radio waves could bounce several times between the earth and the sky, but it was also a disadvantage, because it was complicated to understand whether the source was coming from the earth or the sky.

Therefore, you were likely to receive communications that were either

- in an unknown language, in this case, Russian, or
- extremely distorted, in which, at best, a few words might be comprehensible.

Under these circumstances, it would be easy to misinterpret what you heard.

Such a situation triggers the phenomenon known as pareidolia, in which our brains interpret random noises as known words. Russian-speaking people, listening to these recordings, said that it was very difficult to understand what they heard, with the only clear words coming from radio amateurs, who spoke about more than space communications.

These radio listening stations wanted to satisfy the thirst for news of the latest space enterprises and gave an optimistic interpretation of ambiguous radio transmissions. The Judica Cordiglia brothers, youngsters who were little more than 18 years old at the time, found their house invaded by journalists from all over the world, who hung on their every word and wanted to know any news the brothers could pass on. On the Soviet side, there was always nothing: under this strong psychological pressure, it was enough to receive a few words and end up interpreting them in the wrong way.

Until 1995, this investigation could not have been carried out, because, from the USSR side, there was a wall of silence, and one could only correlate the first-hand information with what was known from the western side. After 1995, there were still problems with language barriers: a lot of the alleged evidence of the lost cosmonauts was in Italian, which you then had to compare with the original Russian sources, which, in turn, required that you be able to speak Russian as well.

And neither Italian nor Russian are simple languages.

Among other things, the Judica Cordiglia brothers decided not to rely on native interpreters in their attempts to interpret the alleged words spoken in the recordings of distressed cosmonauts. In their book, they explain why: when they tried to get in touch with native translators in Turin, sending them the communications, the translations that were returned had very few words transcribed. Then, they got in touch with a German lady who taught Russian at a language school in Turin: this resulted in far more complete transcriptions. Nowadays, there are software programs that allow us to slow down communications while maintaining their pitch, so that the speech you hear sounds quite natural, just much slower; these software programs also allow us to go back and forth in a much easier way than was possible then with the use of tape.

But then, if you play this complicated communication to a person who has been told: "This is a cosmonaut speaking," they will end up hearing words associated with cosmonautics even if what they're listening to is only incomprehensible sounds.

The US engineer and writer James Oberg made the exact same point in his own writings.

2

Yuri Alekseyevich Gagarin

Abstract The extraordinary life of Yuri Gagarin, called the "prince of the cosmos," is detailed. It starts from his triumphal flight and what happened afterwards, then it tells of Gagarin's attempts to save a colleague by preventing the so-called "flying coffin" from being sent into orbit; then, some secrets are reported, both in regard to his flight and to other missions, that were only later revealed. Finally, we examine the disappearance of Yuri himself, at thirty-four years of age, on a plane, and the mystery that still surrounds it.

2.1 Life

With a carpenter father and peasant mother, Yuri Gagarin was a man of the people, the second son of Alexej Ivanovich and Anna Timofeyevna. His mother was also an avid reader.

The family was large and lively: there was the eldest, Valentin, born in 1924, his sister Zoya, in 1927, then Yuri, who came to light on March 9, 1934, and finally, little Boris, in 1936.

It was a quiet life in the village of Klushino, one hundred and sixty kilometers south of Moscow. Yuri's upbringing was based on very solid principles: respect, education, honesty, hard work and an antipathy towards servility.

As a child, Yuri was a curious little rascal, spending his nights watching the starry sky with his uncle Pavel, who frequently joined the family.

But one day, the war came. And the men of the village left to go to the front.

Regrettably, it wasn't that long before the front came to them. As it turned out, the small peasant village was right in the middle of the route of attack of the German tanks.

So it was that Yuri, for the first time, saw what he called "flying men." Everything happened in the sky, and after a duel, two planes with a red star painted on their sides were shot down. The village ran to help the two pilots, one of whose plane had been completely destroyed.

The children offered to escort these two beings from another world, who chose to stay close to their winged steeds that night.

The two pilots left the next day, but in their place came the enemy armies, those with the hooked cross.

The family was thrown out of their house. Boris almost died when a Nazi hanged him from a tree with his scarf. Fortunately, his parents arrived in time to revive him.

In 1943 came the turning point of the war. Hitler's armies turned around. In full retreat, the Nazi army abducted and took away Valentin and Zoya Gagarin, Yuri's brother and sister. Fortunately, they both managed to escape from the concentration camp where they had been imprisoned. They finished the war as part of the Russian army.

After 1945, life resumed in the small village little by little.

A curious detail: the only book that was left for the teacher of the local school to use was the *Manual of Infantry*, and thus it was with that book that she was able to teach reading and writing to her students.

The Gagarin family moved to Gzhatsk, and Yuri began to assert himself at school, where he excelled in mathematics. He joined the Young Pioneers and became interested in physics. He graduated after six years of school. His parents wanted him to do the 7th year as well, which would have been equivalent to a middle school diploma, but he decided that he preferred to get a qualified education instead, attending a night school for young workers and getting a job as a specialist foundry/printer.

He graduated in 1954, at the age of twenty. The time spent in the institute allowed him to come into contact with the writings of E. K. Tsiolkovsky, the father of cosmonautics.

At the beginning of 1955, Yuri made a life-changing decision. He decided to enroll in an evening flight school at the Aeroclub (DOSAAF USSR). There, he had his first experience with skydiving and flying.

It was a challenging time, culminating in his baptism in the air. Yuri was a second officer at the time, on a Yak 18, from which he then jumped with a parachute. The experience remained in his blood.

Shortly afterwards, Yuri made his first solo flight. He would later say that:

Only music could have given expression to the joy of my flight.

Having become proficient with the stick, he decided to go further and join the Soviet Air Force.

In the fall of 1955, his new destination was Orenburg.

While in Orenburg, Gagarin studied at the aviation school named after Voroshilov. It is said that he was almost expelled from the school because he had trouble with landings and could not pass the final test. Eventually, it was noticed that, due to his short height, he did not have an adequate field of vision when landing the plane, which affected his ability to land. He was subsequently given a pillow to sit on, and was thus able to pass the test and graduate.

In January of the following year, Cadet Yuri Gagarin took the oath to the People, the Communist Party and the State. He immediately began to fly MiGs. Day by day, his passion for airplanes grew.

Then, Yuri fell in love. Her name was Valentina Gorjacheva, she loved theater, reading and skating, and she reflected all of the family and moral values under which Yuri had been raised.

The two met at a dance party and instantly liked each other.

In addition to his love for Valya, as his friends called his girlfriend Valentina, Yuri also continued to love MiGs and his studies.

But being a good pilot was not enough for him. He also wanted to become an engineer.

1957 was a great year, for both Russia and for Gagarin:

- Yuri brilliantly passed his final exams at the aviation school and was awarded a certificate with distinction;
- the Soviet Union launched Sputnik 1 into space;
- Valya accepted his marriage proposal, and the couple subsequently married.

It was the feat with Sputnik 1 that pushed Yuri to finish his studies. He attacked them with even more enthusiasm.

There was already talk of sending men into space, and this caused Yuri to dream.

He greatly desired one day to be among those people.

He greatly desired to be the one chosen.

In the meantime, Yuri's path was at a crossroads: should he accept the offer to be an instructor at the school in Orenburg, or go to the southern airfields in Ukraine?

He chose a third, more difficult route: the Murmansk Oblast, in the far north, 300 km beyond the Arctic Circle. At first, he went alone, accompanied by two friends who were aviators like him. Valya had stayed at home to prepare for the exams for her diploma in laboratory technology.

She joined Yuri in August 1958, at a time when he was displaying increasing mastery of refined flying techniques. Valya never warmed to Murmansk, particularly with its horrible climate.

Fortunately, Yuri became friends with the squadron's vice-commander, who was also married, with a daughter. The two families began to hang out and participate in parties and choral concerts organized by the wife of the senior officer.

But the space race had begun, with all of its wonders. After Sputnik 1 came number 2 and number 3. Then came the launch of Lunik 1.

It was on! And it wasn't long before the race reached Yuri's airfield in that remote corner of the world. Government men were sent to the various bases to hold individual interviews with each pilot; in the end, those selected were sent to Moscow, to the military hospital, for further tests.

Yuri was among them. It was his first victory!

In the capital, he was subjected to extremely selective tests: aptitude tests, psychophysical exercises and examinations, mathematical problems to be solved.

The tests went well. He passed. But he was not told why he had been subjected to such a relentless testing regimen. So, while waiting for the second call, Gagarin was sent back to his base of operations. He was back on the ice, his heart pounding with hope.

The second selection process was even tougher than the first.

There were aptitude and physical tests under extreme conditions, then behavioral tests in potentially stressful and dangerous situations. Finally, the reason for so much suffering became clear: Mother Russia was selecting her best men, both physically and mentally, to send them into space. And from over two thousand, two hundred candidates, only twenty men were selected. Twenty.

Yuri was among them.

Having become part of that elite group, Yuri was transferred, with his family, to a new location near Moscow. His wife Valya didn't mind leaving the frozen wasteland forever, but she soon noticed that Yuri was changing. He spoke little about his new assignments, was elusive and seemed to have other things on his mind.

Yuri now knew he was among the men preparing for space! And he wanted to get there first! He had to make it!

Thus, he responded to the hard work that all candidates for astronaut positions were forced to do with his usual courage. The tests were terrible, and dangerous. In fact, in one of them, a colleague of Gagarin's, Valentin Bondarenko, was burned alive.

On January 25, 1961, Gagarin received the confirmation he had been waiting for.

Only six men would compete for the flight into space.

And he was among them.

(free from ESA archives)

The other five were Andrijan Nikolayev, Pavlov Popovich, German Titov, Anatolij Kartašov and Valentin Varlamov. Due to a health problem, Anatolij Kartašov was subsequently replaced by Grigorij Nelyubov. Then, it was the turn of Valentin Varlamov, who injured his head and was replaced by Valerij Bykovskij, leaving the final list at: Nikolayev, Popovich, Titov, Nelyubov, Bykovsky and, of course, Gagarin.

Another misfortune, much worse than Varlamov's, befell Valentin Bondarenko (who was not among the first six, but had been among the initial twenty selected) a few days after Gagarin's launch. We will go into that a bit later.

Among the excluded was Alexeij Leonov, one of the candidates with whom Gagarin had become good friends. According to Sergei Korolev, the two best candidates were Leonov and Gagarin. They were the two names held up for the final choice. But in the end, Alexeij Leonov was excluded. He was too tall! At a height of 1.63 m, he was not very suitable for the small size of the Vostok capsule. Gagarin, on the other hand, was barely 1.57. The young Leonov was distressed, in part because he was actually well below the maximum height allowed for a cosmonaut, i.e., 1.70 m. But as many sources record, the choice was "among the shortest" (meaning that he was short, just not short enough). However, Leonov's worry was short-lived. In 1965, he would have the honor of being the first man to leave a space capsule for a space walk.

A few weeks before the launch, on March 7, 1961, Gagarin's second daughter, Galya, was born. (His firstborn, Melena Lenochka, had been born in 1959.) The newborn was able to enjoy the cuddles of her father for a few days: Yuri had to go to Baikonur to attend the launch of the shuttle Korabl-Sputnik, which was host to yet another dog on board, named Stellina by Yuri. It was to be the last flight before the launch of a man into space.

Gagarin, Titov and Nelyubov were the three selected for the great event in the cosmos. It must be said that Korolev did not like Titov, and he may have had good reason. Titov was proud, sustained [?] and cold, even with friends. In other words, the complete opposite of Gagarin, the prototype of the true Soviet man. In fact, Yuri came from a family of peasants and had been Russian for generations, while Titov was the son of a rural teacher, in short—for those times—of an intellectual.

The Commission for the nomination of the first astronaut decided that Gagarin would be the first Soviet cosmonaut, Titov the reserve and Nelyubov the reserve of the reserve.

With a spread of vodka, oranges and various foods, a big party was held in early April to celebrate Yuri's appointment.

2.2 Those Excluded from the First Flight

There are a number of cosmonauts who could have ended up blowing Gagarin away as the "first man in Space." Valeria Paikova of *Russia Beyond* tells us about some of them, not all of whom were in the top group.

Gagarin had had heavy training and respectable opponents. But in the end, he won. Who were the others, and what happened to them? Some managed to continue flying, even gaining their own bit of glory, while others suffered more painful, if not more dramatic, fates.

Those who applied to be the first man in space obviously had to undergo rigorous testing and exhausting training. Twenty aspiring cosmonauts performed real feats of physical training, under the challenge of testing the limits of Soviet pilots, including in regard to mental preparedness and cardiovascular endurance.

Potential candidates were identified from among fighter pilots who already had experience with G-strength similar to that expected during a space flight. The main criterion for selection was the physical condition of the cosmonauts.

The father of the Soviet space program, Sergei Korolev, believed that an ideal candidate should be about 30 years old, no taller than 1.70 m and about 70 kw.

Here are five of the "competitors" who aspired to the top spot.

2.2.1 German Titov (1935–2000)

The individual who had the best chance of success was Titov. In his book *The Way to the Cosmos* (*Дорога в космос*; *Doróga v kosmos*) Gagarin states that:

> He was as well trained as I was and was probably capable of doing even more. Maybe they didn't send him on the first flight because they were leaving him for the second one, which was going to be more challenging.

In fact, a short time later, it was Titov's turn: between the 6th and 7th of August 1961, also launched from Baikonur, Titov piloted the Vostok-2 capsule for 17 orbits and spent a total of 1 day, 1 h and 18 min in space. Gagarin had only flown for 108 min, and the next launch was scheduled to last even longer, according to scientists. At that point, Titov, just 25 years old, earned a place in history as the second man in orbit and the first to stay in space for longer than 24 h.

2.2.2 Boris Volynov (1934–)

Boris Volynov knew how to wait, as most astronauts or cosmonauts do. And now, in 2021, he is the only surviving male of the first group of Soviet cosmonauts.

From an early age, Volynov's dream was to fly. He was promoted to join the first crew of cosmonauts on March 7, 1960, at the same time that Gagarin was selected.

He engaged in training with the cosmonauts of Vostok 3 and Vostok 4 and was a reserve crew member for Valery Bykovsky, who flew into space aboard Vostok 5, in June 1963. It seemed that it was finally his turn in 1965, when he was appointed commander of Voskhod 3. But in 1966, the mission was cancelled. In its place was conducted the 22-day (unmanned) test flight of Kosmos 110, with two dogs on board, Veterok and Ugoljok.

His moment finally came in 1969, when Volynov spent three days aboard Soyuz 5, becoming the first Jewish cosmonaut in space. After seven years of waiting, his next flight took place in 1976. That mission, the Soyuz 21, lasted 49 days, 6 h and 23 min.

2.2.3 Mars Rafikov (1933–2000)

Rafikov was born in Kyrgyzstan to a Tatar family. In the early 1950s, he earned his pilot's diploma from the local military aviation school and began serving in an air defense unit.

At the age of 26, Rafikov, who, in what was perhaps a mark of destiny, was called "Mars," was chosen for the first astronaut corps, together with Gagarin. So, Rafikov also intensively trained for the launch aboard Vostok 1, and seemed to be among the best. The other members of the group were sure that Mars would be among the first to go into space.

But things went wrong: according to him, it was because of his emotional life. Separated from his wife, he was preparing for divorce when his bosses became interested in his family status. He was ordered to save his marriage for the sake of his image, but he refused.

In 1962, Mars was excluded from the group. Officially, the reason given was that he had left his unit without permission. But he always thought he had been discarded because he had refused to bend to the orders of his superiors, not wanting to lie by pretending that his marriage had not fallen apart.

At that point, the man went back to being a fighter pilot. In the 1980s, he worked in Afghanistan and was awarded the Order of the Red Star.

2.2.4 Valentin Bondarenko (1937–1961)

Bondarenko was born in Kharkov, Ukraine, where, during World War II, many battles between the Nazis and the Red Army took place. Having survived the war, he enlisted in the Soviet Air Force in 1954, at the Voroshilovgrad School. He completed his training at the Flight School of Krasnodar, from which he left with the rank of Lieutenant, and was assigned to the Baltic Military District: it was 1957, the same year of the flight of the first artificial satellite, Sputnik.

Then, he was chosen as a cosmonaut for the flight of the Vostok Program, in the same group as Gagarin and Titov. But nothing was heard of him in the West or the USSR until 1980. Only then did the first news of his dramatic death began to spill out. Not to mention the fact that an attempt had been made to erase him from history.

On March 7, 1960, Valentin Bondarenko was included in the first group of cosmonauts, along with nineteen other pilots: with him were, among others, Yuri Gagarin, German Titov, Alexei Leonov and Vladimir Komarov. None of them, at least during the selective phases, knew what they had been selected for: only when they met Korolev did they understand that they were writing history. It wasn't just a matter of beating the United States of America, it was a challenge to conquer the last frontier of humankind: space.

Bondarenko was valued for his athletic ability and personality. He worked very hard to prove his value.

March 23, 1961, was the third day of a two-week exercise. Part of the cosmonauts' routine training took place in a special isolation chamber, the so-called "chamber of silence." It was like a monk's cell, with basic sanitation, a narrow bed, and a table and seat similar to what they would have in the Vostok capsule. The cosmonauts had to spend 10 days there, and the little room was pressurized to mimic the environment in space, with a very high concentration of oxygen. While in the chamber, the cosmonaut often had to change the many sensors and electrodes that doctors had applied to his body to monitor his vital parameters. Bondarenko was about to leave the chamber—he had just received the green light to remove the biomedical sensors from his body—when he made a fatal mistake. The cosmonaut wiped off the adhesive with a cotton pad soaked in denatured alcohol. On impulse, he tossed it away, with the swab landing right on a hot plate on the stove. A fire broke out, spreading extremely quickly due to the very high concentration of oxygen. Immediately, the flames rose. Unfortunately, it took the technicians almost half an hour to open the pressure chamber: they had to wait for the internal pressure to equal the external pressure. With third-degree burns over ninety percent of his body, Valentin Bondarenko was transported to a Moscow hospital. As he was being rushed to Moscow's Botkin Hospital, he still managed to whisper, "I'm sorry." He remained conscious until the end, dying many hours after the tragedy, eight hours according to some sources, sixteen according to others. The accident was kept secret for a long time. The name of the dead cosmonaut was never mentioned in the official chronicles, and he was also removed from the group photos of the first crew.

At his bedside, in addition to his wife Anya, employed at the Soviet space center, and his son Alexander (who would go on to become a military pilot), there was also one of his best friends, Yuri Gagarin. Bondarenko's death, however, had to remain secret: the KGB decided to erase his presence from every document and every photograph of the first group of selected cosmonauts. Even their attempt to honor him, giving him the Order of the Red Star, awarded posthumously on June 17, 1961, was done in secret. His name would not even appear on the plaque that, on August 2, 1971, the Apollo 15 astronauts deposited on the lunar surface in memory of those who paid with their lives for the benefit of space exploration: simply put, the name of Valentin Bondarenko, for a time, disappeared from history.

James Oberg, in his book *Red Star in Orbit*, also contended that the Soviet government erased the image of cosmonaut Bondarenko from an official 1961 photograph of the first six selected cosmonauts, while British researcher Rex Hall showed that five people had been erased from an earlier group photograph. Clumsy attempts were later made to further curate the historical photographs, with the inclusion of imaginative but non-existent photographic details to explain the absence of the original group members.

2.2.5   Grigory Nelyubov (1934–1966)

Of the twenty men who formed the "Pervyj otrjad kosmonavtov CCCP," the historic "First Cosmonaut Detachment of the USSR" put together between March and June 1960, one died, and seven were excluded: three for health reasons and four for disciplinary reasons. Among these was Nelyubov, one of the first to be admitted to the group. He was Gagarin's second replacement and was prepared to become the third Soviet cosmonaut in space. His flight was scheduled for the fall of 1961.

When the commission met to decide who would be the first man in space, it is true that the three candidates were Nelyubov, Gagarin and Titov, but the results were so similar that the race continued until the last useful week. In the end, it was decided that Gagarin would pilot the first Vostok, Titov and Nelyubov the following ones. In fact, Titov would have piloted the Vostok-2, while Nelyubov should have been on the Vostok-3, but he was replaced, and was only the second choice for the Vostok-2. His colleagues were sure that, in the future, someone as good as Nelyubov would surely end up piloting a Vostok, and so were the instructors.

But it didn't happen.

Why? It had nothing to do with his skills as a cosmonaut, but rather with his character flaws. An episode demonstrates this. In 1960, Kamanin wanted an internal mini-vote to measure the mood of the group. There were twenty people, and each had to predict who would fly first on Vostok. The vote was anonymous.

Gagarin received 17 votes and Nelyubov none. The reason? Gagarin was a modest and easygoing person; not only that, when he was wrong, he apologized. Nelyubov, in contrast, was vain: he emphasized his best results, but blamed others if he failed, an attitude that the USSR did not like. Individualism was frowned upon. Cosmonauts were prominent figures, they were the symbol of Soviet values.

Nelyubov was considered a top level cosmonaut and a very good pilot, with an exceptional mind and a physique able to withstand accelerations of 10 G for tens of seconds without losing consciousness, as happened to his companions. What's more, he was charming and a great talker, with a strong personality.

But if any of the cosmonauts got out of line in terms of discipline, responsibility always fell on Kamanin, and Nelyubov's shenanigans were the sort that could not be tolerated. But, on the other hand, the USSR couldn't afford to give up such a good cosmonaut, so his flight was only delayed.

Until he took one step too far.

In the spring of 1963, Nelyubov and two colleagues (Filat'ev and Anikeev) went out for a drink. This in itself was already an impropriety, and in addition, the three were wearing military uniforms at the time. They ate and drank (a lot) in a small restaurant. They began to annoy the other customers, until the manager of the restaurant asked them to leave, over and over again. Finally, fed up, he called the police. At this point, the cosmonauts (drunk, but not to the point of thinking that Kamanin would be angry) fled, but were subsequently stopped by a patrol car. Seeing the uniforms, the policemen decided only to give them a verbal lecture; after all, they were colleagues. Then, one of them asked for the cosmonauts' documents, which the three did not have on them. The policemen took them in to the station for identification. Afraid that a report would tarnish their record and get them kicked out of the space program, the three tried to run away, starting a fight with and beating up the arresting officers. But it wasn't long before they found themselves back at the police station. The captain called the training center, which confirmed the identity of the three, promising at the same time that there would be punishment for their behavior. However, the captain was apparently not eager to get them into trouble, and so he decided to let them go without a report. His proposal was: apologize and let's end it here. Filat'ev and Anikeev willingly apologized. Nelyubov, however, stubborn as ever, would not. He told them: I am a cosmonaut, the best, I have many connections, it is you who should apologize. Not even a saint would have let him get away with that. The captain sent him home, but promised that if he did not apologize by the next day, he would be reported directly to Kamanin. The next day, his comrades spoke to him, appealing to his common sense, but Nelyubov would not back down! So, the famous police report reached Kamanin, who became very angry. He gathered the cosmonauts and gave them a very harsh lecture.

In doing so, he kept in mind the misdeeds of Gagarin and Titov.

It is worth noting that, having become Heroes of the Soviet Union, Gagarin and Titov began acting like prima donnas. As a living legend, Gagarin was always surrounded by women, and did hot hesitate to betray his wife when he could, after which the party was obliged to cover up his shames so as to keep his public image clean. Titov, for his part, drove sports cars at insane speeds, sometimes drunk out of his mind, and caused a number of accidents. He once ran over and killed a woman, and paid off a witness to avoid being convicted, thus forcing the party to bribe the victim's family to ensure their silence.

With these memories in mind, Kamanin spoke harshly to his men, after which he called for a vote by a show of hands so that it would be clear where the group stood concerning the expulsion of the three. The others were not actually convinced. To them, Nelyubov was an irreplaceable member of the team. However, they did not dare to oppose Kamanin, and thus voted in favor of the expulsion. From one day to the next, Nelyubov went from being an excellent cosmonaut to a great nuisance. He was transferred to the small village of Krenovo, where he resumed his piloting duties.

Neither Nelyubov nor his wife could tolerate living off the grid. He made several requests to be readmitted to the space program, but never received a response. He also tried traveling to Moscow so as to speak directly with Kamanin, but he was not received. He then tried to be hired as a test pilot for the new MiG-21; he passed all the selections and was called to Moscow, but a few days later, he was told that his position had been cancelled. The air force leadership had spoken to Nelyubov's former comrades. They had learned that the man was an excellent pilot, but unreliable. They decided to give his assignment to someone else. At the same time, some photos related to the space program started to appear in the newspapers, and Nelyubov saw, with horror, that he had been erased everywhere. This led him to fall into depression. He would frink heavily, and his wife was forced to lock him in their house to give him time to sober up. This situation continued until the day that he ventured out of his house for one last time. "Russia Beyond" states that:

> He made several attempts to get back in [to the space program], partly because Korolev still had faith in him. But after the death of his mentor in 1966, the doors of space closed completely for him. Shortly thereafter, drunk out of his mind, he was run over by a train in a small station in the Russian Far East.

The accident occurred on February 18, 1966, in Krenovo, a remote village north of Vladivostok, on a foggy day. Apparently, the engineer, having noticed the man on the rails at the last moment, tried to brake, but failed. The body was completely mangled. It belonged to a man who used to stay out late in bars, where he would tell the long story detailing how he was the one who should go into space, and not Gagarin. In the report, the police said that the man, later identified as Grigory Nelyubov, had, due to cold and drunkenness, carelessly crossed the tracks, and had not seen the train coming. An accident. Almost no one attended the funeral. Thus, Nelyubov was buried in a common grave, in the small cemetery in Krenovo. His grave remained forgotten for 20 years until, in 1986, the story came to light. A news agency organized a collection and, today, there is a small marble monument on his grave that remembers him as a pilot and cosmonaut of the Soviet space program.

*From Russia Beyond https://lit.rbth.com/scienza-e-tech/85316-perch%C3%A9-gli-ast
ronauti-della-nasa*

This is one of the few photos showing Nelyubov in the group of cosmonauts. He is the third from the right, in the second row.

But, as far as all other documents from the period are concerned, he had
already disappeared three years before his death, erased from (almost) every
photo, document and publication.

Disappeared from group photos, films, records. And from the famous
holiday photos that cosmonauts and instructors were known to give each
other.

In May 1961, a few weeks after Gagarin's flight, the cosmonauts and
instructors went on vacation to Sochi, on the Black Sea, to celebrate and rest.
The atmosphere was festive and photos were taken. Years later, the photos
were given to the press and became world famous. One of the most famous
photos captured a group of eight, five people sitting and three standing
behind them. At the center were Gagarin and Korolev; the others would
go down in history as the "Sochi Six," the Soviet Union's most promising
cosmonauts, who would take part in their nation's successes.

But something wasn't quite right. In the back row, there was an empty
space. Additionally, in some other versions of this photo, there is a ladder,
while in others, there is a hedge. Someone who was part of the top group of
Soviet cosmonautics had been eliminated. Who was it? Western intelligence
services got to work and another photo was discovered. One from April 12,
1961. Gagarin is walking toward the launch pad accompanied by another

cosmonaut. The man behind Gagarin was one of the reserve cosmonauts who, in all likelihood, would have led one of the subsequent missions.

Yet, that face does not appear in any other photos. The man never took part in any mission. Despite this, there are other records of him. Some photos show blank spaces, others yellowed spots used to blur his face. Who was he?

And why such an extensive attempt to conceal his identity?

Had there really been secret missions that had failed?

For twenty years, there was no answer.

Until it turned out that it was, indeed, Nelyubov.

2.3 The First Flight

One hundred and eighty-eight minutes in space, eighty-eight of them in orbit:

> To put a man on a multistage rocket and launch it into the gravitational control field of the Moon where passengers could make scientific observations, perhaps land alive, and then return to Earth - all this constitutes a wild dream worthy of Jules Verne. I am reckless enough to argue that such a man-made trip will never happen despite all future advances. (Lee de Forest [1873-1961] scientist, inventor, filmmaker)

The flight was imminent.

And, finally, the big day arrived: on April 11, the R-7 rocket was brought to the launch pad.

Gagarin and Titov spent a quiet day. In the evening, they played a game of billiards. Then, they went to sleep, and each stage of sleep was monitored by the scientists at the space center.

The morning of April 12, 1961, was to be written in history.

Gagarin was awakened at 5:30 a.m., performed his usual exercises and washed up.

For breakfast, he had a "space menu": minced meat, blackberry jam and coffee.

Then, Gagarin and Titov began their preparations for the mission: they wore warm, light blue undersuits, over which were orange protective suits equipped with a pressurization, ventilation and power system.

After dressing, Gagarin, Titov and Nelyubov (without a spacesuit) were taken by bus to the rendezvous with the R-7 rocket.

According to space historian Asif Azam Siddiqi:

Engineer Sergei Korolev, supervisor of the Vostok 1 mission, was so agitated on the morning of April 12, 1961, that he had to take a heart pill.

Gagarin, on the other hand, was calm, at least in appearance.

(In contrast, a half an hour after the launch, his pulse would register sixty-four beats per minute.)

On the way to the launch pad, Gagarin stopped to urinate on the back wheel of the bus carrying him.

Since then, that gesture has become an obligatory and propitiatory ritual for all Soyuz astronauts, at least for the males.

But that was not the only tradition perpetuated in Gagarin's memory: there was also the haircut two days before the launch, drinking a glass of Champagne on the morning of departure, the signature on the door of the hotel room before leaving to go to the ramp and deliberately not attending the transport and positioning of the rockets and the spacecraft itself.

Finally, Yuri entered the capsule.

He did the final checks, quickly going over the emergency procedures and the unlock code for the manual controls.

Then, the pod door closed.

To calm himself down during the wait, half-serious and half-joking, Yuri asked for some music. And thus music was played.

At 8:51 a.m., the music was interrupted by Korolev's voice.

Then, there were communications in rapid succession, followed by *"launch key in starting position"*... *"air released. Ignition."*

How was it possible to schedule the flight for that date? The fact was that they had no choice but to hurry. A rumor had reached the Soviet leadership: the Americans were planning to send their first man into space before the USSR. The Soviets could not allow this.

So, the Vostok-1 spacecraft was hastily designed.

Some systems were not sufficiently reliable.

Yet, both chief designer Korolev and young pilot Gagarin decided to take a chance.

On April 12, 1961, the Vostok-1 spacecraft departed from Baikonur.

Gagarin sat on board, and the whole world held its breath.

The first man in space spent one hour and forty-eight minutes in space and returned to Earth victorious, despite there being no shortage of emergencies during the landing.

Gagarin immediately became an international celebrity, indeed, a hero of exceptional greatness. His charismatic smile became a symbol of triumph and the achievements of humanity.

How did the adventure begin?

"Here we go!!!" yelled Yuri, and the rocket lifted off from the earth at 9:07 a.m.

Then, the Semyorka launcher took off so as to launch the small Vostok 1 (it was almost claustrophobic) into low orbit with Gagarin inside.

The man made a complete orbit around the Earth, with a maximum altitude (apogee) of three hundred and two kilometers and a minimum (perigee) of one hundred and seventy-five.

The speed was astounding for the time: twenty-seven thousand, four hundred kilometers per hour.

As commander, Gagarin had chosen his own code name: кедр ("kedr", "cedar").

Within nine minutes of the launch, Vostok 1 (meaning East 1) was in Earth's orbit.

Yuri remained in contact with the ground, broadcasting:

I feel good, the flight is going smoothly!

Then, he uttered the famous phrase:

From up here, the Earth is beautiful, without borders or boundaries.

Looking down from the spacecraft, the first human being to do so, he said that:

The Earth is blue [...] How wonderful. It's incredible!

The spacecraft, after twenty minutes, carried itself to the Arctic Circle, then to the North Pacific. Yuri exchanged a few jokes with his friend Leonov, who, until the last moment, had not known who would go into orbit first, between Gagarin and Titov.

Then, Gagarin found himself over the South Atlantic:

The Russians have a man in space and the United States is asleep.

The US newspapers were full of bitter commentary the next day, while the whole world continued its vigil.

After seventy-nine minutes, the retrorockets came on and the shuttle automatically assumed the correct orientation for the return course. Yuri flew a full eighty-eight minutes around the earth, spending this time as a passenger. In fact, the controls were operated by a computer on the ground. They could be freed for operation on board through use of a special key, but they were usually locked. For the duration of the mission, four ground radio stations transmitted music, interspersed every thirty seconds with a call message in Morse code: this allowed Gagarin to choose the best frequency with which to communicate.

But this all happened within the context of a full Cold War. The Americans wanted proof that the Soviets had actually sent the first man into space. It had been rumored for some time, but the US could not be sure.

Or was it just propaganda?

For this reason, even before the launch, the US National Security Agency had designed and built special stations capable of intercepting Russian communications. One of these was located in Shemya, in the Aleutian archipelago (Alaska), and was able to capture communications between the cosmonaut and the ground base, demodulating the video transmission and allowing the Americans to see the images of Gagarin inside Vostok. This had already happened with the two previous launches of Vostok, which had had on board, respectively, a dog and a dummy. So, only fifty-eight minutes after the launch, the US military leaders had their confirmation: the Soviet Union was being serious.

Then, the spacecraft slammed on its brakes, ignited its retrorockets and returned Yuri to Earth's atmosphere. The flight ended at 10:55[1] a.m. Moscow time, in a field south of the town of Engels (Saratov Oblast), a little further west than planned. The cosmonaut was ejected from the cockpit and parachuted to the ground. In the official documents, it was said that he had landed while still onboard the capsule.

The truth was kept secret for two reasons. First, they did not want to reveal that the USSR did not yet have the technology for the soft landing of a spacecraft. Second, they did not want to invalidate the approval of the record by the FAI (Fédération Aéronautique Internationale), which required the pilot to remain on board until landing.

Many sources say that this was kept secret until 1971.

But by 1961, the road to space was open.

Upon his return to Moscow, thousands of people gave Gagarin the honors due to a Hero of the Soviet Union. Newspapers around the world published photos of the Major, with his robust physique and open boyish face. "Major" because he had left for space with the rank of Lieutenant, but Nikita Khrushchev had promoted him during the flight.

---

[1] 10:55 a.m. was the officially reported time, however, the actual time was 10:53 a m. After the documents were declassified, this came up and was widely discussed among specialists and amateurs.

In the meantime, Yuri's parents were tracked down and informed: their son had become a hero. TASS communiqués were initially cryptic, but they became more and more explicit and triumphant.

The whole story of Yuri Gagarin stands as a triumph of Russian propaganda, which was very clever at concealing any real or supposed truth in favour of the only fact that really mattered: the fact that a man had gone into space. A Russian man. A great deal of material relating to the first flight remained classified until 2011.

At the end of the last century, Russian children could visit the VDNH exhibition (an exhibition of all of the achievements of the national economy of the USSR), where, in the Cosmonautics Pavilion, there were replicas of Sputnik and Vostok and all the initial rockets. It was not as accessible to visitors as the exhibitions in the Kennedy Centre in the US (you could not actually step inside any of the vehicles), but it was still very impressive.

Let's take a look at what the Vostok 1 spacecraft looked like.

It was 4.4 m high and weighed a total of 4.7 tonnes.

It consisted of two parts:

- a spherical habitable module, which housed the cosmonaut, and
- a service module that housed the on-board instrumentation, the retro-rockets needed to brake and return the probe to Earth, and the sixteen tanks containing oxygen and nitrogen.

The habitable capsule had three portholes, an optical viewer to be oriented by hand, a camera, instrumentation to measure pressure, temperature and orbital parameters, a hatch and an ejector seat, and was about as long as the cockpit of a Fiat 500 of the time.

Re-entry was not all sunshine and rainbows, in fact, some problems arose:

- The heat shield risked being burnt out, and
- The spacecraft lost its buoyancy and began to spin around, almost causing Gagarin to lose consciousness.

At seven thousand feet, the door opened and Gagarin was ejected. At four thousand, he separated from his seat. Then, a large parachute opened.

With that great blanket over his head, Yuri soon found himself in a soft field in his beloved Russia. He was found and embraced by a peasant woman named Anna Taktarova.

The wife of the local forester of the farm in the village of Smelovka (Saratov region), Taktarova had no idea that, on that April day, she would experience her golden moment, one to be told in front of the fireplace in the evening. But it happened. A strange man fell from above. Hanging on to something that could have been a large blanket or a large umbrella, he swooped down onto a field near her house. The two embraced as if they had known each other all their lives. It seemed to Anna Taktarova, a peasant woman from the vast nation of Russia, that this was the best way to greet the alien who had fallen from the sky.

Yuri found the field to be more beautiful and soft than he had ever imagined a field could be. He took off his helmet and went back to breathing real air. When the woman ran towards him, the embrace that followed was, for him, the ideal way to be welcomed back to Earth. Safe, sound and in one piece. A scene from the movies. To think that, 108 min earlier, he had been on the launch pad of Baikonur, inside the Vostok 1 capsule, on top of a rocket aimed at the stars.

Before long, the anonymity of the village of Smelovka was no more. Men in grey coats arrived, military men. They took what there was to take and left, only to return later to make sure that they hadn't forgotten anything.

Anna Taktarova had experienced her moment of glory: she had been the first person to welcome back the first man into space. From then on, she would only see him again on newspaper covers or hear him on the radio: he was a Hero of the Soviet Union.

This is the more romantic version. Other sources say that Gagarin had been in danger of landing in the Volga River, but was able to manoeuvre his parachute by pulling on the straps.

In any case, on April 12, 1961, the history of the conquest of the stars was rewritten by the voice of Radio Moscow, which announced, in English as well as Russian, the feat of Major Yuri Gagarin.[2]

Then, the men of the regime arrived and took the hero of the USSR to Kuibishev. Yuri was exhausted, but he spoke to Nikita Khrushchev on the phone. He was greeted triumphantly by colleagues, scientists, men of the regime and all of the others who had participated in writing this page of history. After he underwent a number of medical examinations, the day came to an end. Finally, it was time for the usual game of billiards with Titov.

---

[2] The Russian people were given little information about such a significant event and its many participants. Out in the public sphere and in the school textbooks, there was a very nice but artificially error-free story that the majority of Soviet people neither challenged nor doubted. There was a bit more information circulating in the cosmonautics research and engineering institutes, as well as in cosmonautic and aeronautics clubs where people did their own calculations of the flight characteristics, but even most of that information came from the official releases and the rest was classified.

Yuri was only able to hear his family over the phone. He had to wait until April 14, when he was to return to Moscow, to see them.

This was the beginning of a continuous mob scene for him. Radio, television and newspapers competed for photos and statements from the cosmonaut.

The Hero of the Soviet Union was given some benefits that the Party gave to its best men. Thus, Gagarin could have his own car, a four-room flat in Moscow in a residential area and a dacha in the countryside for weekends. He was appointed US-USSR peace ambassador and embarked on a long world tour that would take him to various parts of the globe.

The indoctrination to which he was subjected by the Party forced Yuri to gloss over both the problems he had in space and the fact that he arrived on earth hanging from a parachute and not in the capsule, as was reported by all the news sources around the world. But it was hard to be a living myth, a myth whom the long arm of the Soviet regime followed, spied on and accompanied at all times.

Pretty soon, Yuri began to change.

2.4   After the Flight

Part of the historiography depicted him as a drunkard who ran after every skirt. Regardless of how true this was, the regime's historians had to prevent the myth from being undermined at all costs.

According to some sources, Yuri apparently eventually lost control.

One evening, in a drunken stupor and captivated by his interest in a nurse who was working at a holiday resort for powerful Russians, he suffered a cut on his head in his attempt to avoid being caught red-handed in the company of the girl.

Khrushchev wanted to present a more relaxed image of Russia and its best men, free of any stain or sin. In presentation, the Soviets were not at all moralistic. But that was only a facade. Gagarin recovered from his injury and the doctors who treated him were praised. Thus, the regime closed the incident at home.

But relations between Yuri and his wife deteriorated.

For her, being the companion of a legend came at too high a cost.

In fact, the legendary man began to feel uncomfortable with the high Soviet hierarchy: they no longer served the interests of the people, they were only focused on selling Russia's image and enjoying various privileges and powers.

In 1963, Yuri resumed his studies.

The new generation of cosmonauts was fierce.

It was not enough to be a skilled pilot in order to go to space.

In 1964, Yuri went to the prestigious Zukovsky Academy in Moscow, where he discussed a futuristic thesis of incredible scope: he was thinking of using a spacecraft equipped with wings that could take off like a rocket, go into orbit, and then return to Earth like a plane, piloted by astronauts. Basically, Yuri was thinking of something like the shuttle, a full 20 years earlier.

But the general situation was taking a turn for the worse. While it was true that the Soviet space programme had gone from one success to the next, it was also true that it was running out of steam, and that there were enormous problems to be concealed. For the sake of political necessity and propaganda, delays in the conduct of the space programme were not permitted. It was now common practice to overlook the safety of the astronauts.

Things got even worse when Khrushchev was succeeded by Brezhnev. Gagarin's public appearances decreased, and he was being watched even more closely by the secret service, who had replaced all of his usual bodyguards because he had become friends with all of them.

Then, in the midst of a flurry of activity, including the exploits of Valentina Tereschkova, the first female astronaut, Leonov with his spacewalk and the three astronauts aboard Voskhod 1, Sergei Korolev fell seriously ill.

In January 1966, the chief designer of the internal Soviet space programme was admitted for an operation on his intestines. His condition was bad, and his body turned out to be too weakened to withstand the operation. He had been undermined by his imprisonment in Siberia during Stalin's rule and the relentless pace of work to keep up with the programme. The Soviet Von Braun's life was cut short by a haemorrhage that the doctors were unable to stop.

But, as it happened, two days before the fatal operation, Korolev had managed to recount his hardships under the Stalinist dictatorship to Yuri, who was deeply affected by the story.

After this, Gagarin's disappointment with the new Soviet nomenklatura grew. He decided to return to space; indeed, he made a vow to take Korolev's ashes to the Moon. He threw himself headlong into the task, with the same determination that had distinguished him in his studies.

2.5   The Flying Coffin

Gagarin achieved his first victory when he was appointed as a replacement cosmonaut for the upcoming Soyuz 1 mission.

The mission's first cosmonaut, Vladimir Komarov, was all too aware of that which had become terribly apparent: the flight aboard the Soyuz would almost certainly be a flight of no return. Gagarin knew it too, as did they all.

In vain, he and his fellow cosmonauts drew up a ten-page document on the mission's shortcomings. In summary, they said that the Soyuz was a flying coffin, and that the authorities should have known this and suspended the flight.

But the space race did not allow for exceptions or delays. Politics entered heavily into it, and thus the Soyuz was "ready to go" near the end of April 1967, with its two hundred and three documented construction flaws. Two hundred and three! Many of those who had worked on the document were dismissed from the Soviet space programme and disgraced.

Gagarin tried in every way possible to cancel the launch, and even attempted to take the place of his friend Komarov. All to no avail. The regime wanted it that way. Thus, on April 23, Komarov was launched on a mission from which he would not return alive!

Gagarin was devastated. He felt betrayed by the country that he had loved so much. The situation in Star City, the Soviet space centre, had also become untenable. The new astronauts, among whom Georgji Beregovoi stood out, represented the most careerist and ruthless expression of the Brezhnev-supported regime, and, above all, they were inextricably tied to the central power. In fact, Beregovoi, who detested Gagarin, was just beginning an unstoppable career. He started with the departure of Soyuz 3 in October 1968, and within four years, he was in charge of Star City.

Mother Russia's stubborn peasant son was left with only the spy microphones that the KGB had placed all over him. Gagarin had become unreliable, a danger to the regime's programmes.

He was still a Hero of Russia, of course, but had become nothing more then an entertainment, a sort of puppet hailed by the world. His chances of flying again or of having a say in the space programmes had been wiped out. Even on the MiGs, Yuri could no longer accrue significant flight hours, and Beregovoi never missed a chance to point out that he himself, as a decorated war hero, was far superior to Gagarin.

But the "Colombus of the Cosmos" still had a fighting spirit.

Yuri turned to an experienced instructor, Vladimir Serjogin, in order to be able to fly fighters again. As we shall see, that choice cost him his life.

2.6   After the Cold War—Declassified Documents Offer More Details About Yuri Gagarin's Flight

The end of the Cold War brought a great deal of information about the Soviet space programme to light. That data was gathered by the best Russian journalists, i.e., those able to track down veterans who were willing to talk. The result was a kind of revisionist history, about "what really happened," rather than "what we thought had happened."

Despite the value of all that work, the main challenge of Soviet space history has always been the problem of archival research. How do you dig through Moscow's archives to get the documents? It's not like the American space programme....

Only since the 1990s has it been possible to visit those archives and get access to Party and government documents. It was not easy, but there have been a number of academics, both professors and university students, who have done a lot of research. The archival documents on the Soviet space programme had become available. And there were a lot of them.

This was both an advantage and a source of confusion.

The Russian archival authorities, for example, have published several collections of documents from the early days of the space programme (all in Russian), which are now commercially available.

But there is some selection bias between what has been included and what has been omitted. Selection bias is, of course, a problem with any published collection of archival documents, but the Russian ones present their own peculiar set of problems.

The wealth of material is surprising.

So much material on the space programme and related fields has been declassified in recent years. These include: interplanetary plans and programmes; detailed lists of technical materials from the American aerospace industry, much coveted by Soviet industrial executives; documents complaining about intelligence leaks at Baikonur; abandoned anti-satellite projects; documents on the massive N-1 Moon programme.

The set of documents on Yuri Gagarin's historic flight revealed the enormous risks associated with that mission. The flight has been extensively documented, in print and online. However, the great Russian declassification of 2011 clarified a lot of things that few people were aware of. For example, as detailed earlier, many websites say that Gagarin was not the first human being in space, that there were "lost cosmonauts" before him. The truth is that nothing is known for sure, in fact, the latest interpretations say that the "lost cosmonauts" were just … dummies!

The text within the documents on Gagarin was classified as "Top Secret." It spoke of his mission, and had been prepared by planners for use by the highest levels of the Soviet government. This five-page report, dated May 9, 1961, less than a month after Gagarin's flight, summarizes all that the engineers knew about the flight. How did Gagarin do? How well did his spacecraft perform?

How did he feel during the flight? What were the critical issues?

And as to now, what can we conclude from the information that has been revealed?

– To begin with, we can dismiss the idea that Gagarin was unwell during the flight. The authors of the paper note that "*Senior Cosmonaut Yu. A. Gagarin: normal health status during insertion of [his] ship into orbit, space flight and return to Earth, maintaining full work capacity during flight and completing flight assignment and observation program.*"

In addition to the paper's comment about an "observation programme," we have an explicit confirmation of the military importance of Gagarin's flight in the next sentence, when the authors note that the flight "opened up new perspectives in the mastery of cosmic space and the use of these objects for defence." Remember that Gagarin's flight was a military one. His spacecraft was a variant derived from a new spy satellite: the "Zenit."

Moreover, the military aware of the operation did not want others to mistake him for a stranger, that is, a saboteur. So, about twenty minutes before departure, it was decided that the letters 'CCCP' or"URSS" would be engraved on the cosmonaut's helmet, although it was the fact that the first man in space exited his craft with untied bootlaces that went down in history.

– Secondly, the "air conditioning" (i.e., the life support system on Vostok) "did not fully match the [design] requirements," meaning that, for Gagarin, the life support was operating right at its limits.
– Thirdly, the "portable emergency reserve," i.e., a package used by cosmonauts to survive (for about three days longer) in the event of a landing in an unexpected area, had not been thoroughly checked. The document notes: after Gagarin had ejected from the capsule, when he was parachuting, "the cable connected to the [portable emergency reserve] snapped," depriving him of these supplies. In other words, if he had landed too far from the target, he would have had to make it without supplies.

– Fourthly, a valve was incorrectly assembled. The paper notes: "This could have led to premature engine shutdown and [failure] of the spacecraft's orbital insertion." Had that occured, one can imagine what would have happened to Gagarin. The best case scenario was an unplanned landing, perhaps in eastern Siberia. The worst case, given all the unknowns, was a fatality. Indeed, both this valve and its operation during orbital insertion did endanger Gagarin's life, but in an unexpected way, as we shall see.

– Fifthly, the shortwave mode for the voice radio system "during the cosmonaut's flight did not provide for normal communications with ground communication stations," which explains the repeated complaints: ground control and Gagarin had difficulty hearing each other, not to mention that the audio quality was poor. Nonetheless, Gagarin logged some vivid real-time impressions on a recorder of his time in orbit. ('*The flight is going wonderfully. The feeling of weightlessness is not a problem, I feel good.... At the edge of the Earth, at the edge of the horizon, there is such a beautiful blue halo that it gets darker the further it is from the Earth...*')

– Sixthly, during the flight, one of the two on-board radar sensors, the one that helped ground control to track the spaceship's coordinates, failed. This meant that the tracking data was spotty during the mission.

– Seventhly, due to "incorrect assembly" at the factory, the spacecraft's main data recorder (a kind of black box) known as "Mir-V1" failed during re-entry and landing. What did this mean? It meant that a lot of critical data about the last part of Gagarin's mission was not recorded, making troubleshooting after the mission all the more difficult.

We also know this: there were a number of other anomalies (in NASA parlance) that endangered the flight, including one that could potentially have killed Gagarin. During the orbit launch, due to the infamous faulty valve, the upper stage engine ran for longer than it should have, putting Gagarin into a much higher orbit than expected: the apogee of the orbit was 327 kms, instead of 230. That is, in the event of a failure of the retrorocket system, Gagarin's ship would not have decayed naturally after about a week or even ten days, the absolute limit of the ship's resources, as planned. Instead, it would not have returned until after a period of 30 days. By that time, all of the air inside the ship would have been exhausted, meaning Yuri would surely be dead. In other words, either the retrorocket worked, or Gagarin was a dead man.

During the actual flight, as soon as orbital insertion occurred, a timer went off. Just sixty-seven minutes later, this timer sent a signal to turn on the retrorocket engine, which did its job and deorbited Gagarin.

In retrospect, the fact that the retrorocket engine ignited as planned is not surprising, since it was one of the elements of the entire spacecraft that had been most tested at the factory.

Interestingly, the entire time he was in space, Gagarin had no idea that he was in the wrong orbit.

A serious problem had occurred when, after being switched on, the retrorocket engine had stopped working after 44 s, i.e., one second before the scheduled shutdown time, due to another faulty valve. This one-second difference meant that Gagarin would have landed 300 kms from the intended point. The lack of a proper cessation also meant that some propellants were causing Gagarin's ship to spin uncontrollably (about 30 degrees per second).

Gagarin, affable as ever, referred to it in his later post-flight report as a "dancing body" as the ship spun madly. He recalled that it was:

> ...head, then feet, head, then feet, spinning rapidly. Everything was spinning. Now I see Africa... next to the horizon, then to the sky... I was wondering what was going on.

The problem was much more serious than expected, as this crazy rotation interrupted the internal programme. This would lead (four to eight seconds after the engine had stopped) to the failure to separate of the two modules that made up the Vostok spacecraft, namely, the spherical descent module (carrying Gagarin) and the conical module, which lacked a heat shield but would—in theory—burn separately, away from Gagarin's capsule.

In his post-flight report, he recalled:

> I waited for separation. But there was no separation.

The fact is that the two objects began to enter the atmosphere while still chained to each other. This was extremely dangerous: some parts of the module had not been designed to survive re-entry, and they could easily have hit Gagarin's capsule.

Fortunately for him, about ten minutes later, the two parts of Vostok separated, at an altitude of about 150–170 kms above the Mediterranean. Lower than expected, but still high enough to save Gagarin's capsule. Yet even then, not everything was safe. For a few seconds, a harness kept the two modules connected as they danced wildly. They were only separated when four steel strips securing the harness itself came off. Luckily!

After experiencing an acceleration of about 10–12 Gs during re-entry, once in the atmosphere, Gagarin was ejected from his capsule at an altitude of about seven kilometres. However, he soon discovered something else: having deployed his large primary parachute, there was a problem with the slightly smaller secondary parachute, which only partially deployed. Having one fully deployed and one partially deployed parachute was a recipe for disaster, but fortunately, it did not adversely affect the descent.

Once again, Yuri was lucky.

Gagarin was busy with other problems: for six minutes, as he descended, he had struggled to open a breathing valve on his spacesuit so that he could breathe atmospheric air. His life was not in danger, but it was extremely uncomfortable.

Gagarin was lucky to have come out alive and only injured. In the duel with the United States, Soviet engineers pushed the boundaries of acceptable risk to the limits. Fortunately for the USSR planners, all went well. Of course, part of this was due to luck. But part of it was also due to the undeniably robust design of the Vostok spacecraft.

The designs of the Vostok project were (relatively) simple and elegant, designed primarily to get a person into orbit and back as quickly and reliably as possible.

The Soviets dropped the truncated cone design (like the one used on NASA's Mercury spacecraft) in favour of a sphere: and, as it turned out, the sphere was the shape most capable of re-entering the Earth's atmosphere in the easiest way.

The problems that Gagarin faced on his mission were not necessarily due to poor design or engineering, but more a combination of haste and poor manufacturing. Let's look at the design of the Vostok spacecraft.

It was made of 241 thermionic valves, over 6,000 transistors, 56 electric motors and 800 relays and switches connected by some 15 kms of cable. In addition, there were 880 plug-in connectors, each (on average) with 850 contact points. In total, 123 organisations, including 36 factories, helped to build the entire Vostok system.

Despite the fact that a large number of the systems were redundant, the cosmonauts still could not have absolute confidence in the spacecraft.

But the Soviet engineers had to design a system that worked even with the parameters pushed to the limit.

That's why the design of the Vostok spacecraft was the result of excellent and robust engineering. No, the problem with Vostok was something else entirely: it had not been sufficiently tested.

Thus, the frenetic pace of the 'space race' sacrificed the practice of thorough testing on the ground in favour of debugging the technology in space.

People aboard spaceships had to take more risks.

Mission managers felt that it was too hard to test systems on the ground (instead of in flight).

Depending on the urgency, sometimes you do something on a mission that you haven't really tested in the field.

Apollo 1 was an example of this.

The question is:

Will it do its job, even if a lot of things go wrong?

In the case of Gagarin, the answer was:

Yes. Because beyond all the problems of the mission, he will always be the first human being in space.

Note that UHF communications with Vostok were maintained from the moment Gagarin entered the capsule until about 23 min after launch. Then, contact was switched to shortwave, from various ground stations. But neither Novosibirsk nor Alma-Ata received any word from Gagarin, while Khabarovsk maintained two-way communication for only four minutes (from 09:53 to 09:57, Moscow time) and Moscow for about one minute (starting at 10:13, Moscow time).

However, the fact remains: Gagarin proved that human beings were capable of flying beyond all expectations. At just 27 years old, he became the first man in history to orbit the Earth and observe it from space. Nikita Khrushchev conferred upon him the Order of Lenin, the highest Soviet honour, and, as mentioned above, he was named a Hero of the Soviet Union.

Gagarin, before his fall from grace, would continue to collaborate in the preparation of other space missions, such as Vostok 6, which, in 1963, led to Valentina Tereškova becoming the first woman in history to fly in space. He also participated in the development of the new Soyuz spacecraft.

As an anecdote, we mention that, prior to the space flight, three messages were recorded "from the first cosmonaut to the Soviet people": only the first was recorded by Gagarin himself, while the other two were made by his replacements, German Titov and Grigory Nelyubov.

2.7   Other Details that Emerged Later: There Was an Emergency Situation Just Before Departure

No one could be sure that everything would go well.

It had previously happened just before the flight of Laika's Vostok-1: an emergency situation had occurred, i.e., the sensor for the air-tightness control of the door had not given the desired signal. With only a short time to go before departure, the problem could have caused a delay in the flight, but the workers and the project director, Oleg Ivanovsky, would not give up. In a matter of minutes, they managed to tighten over thirty screws, adjust the sensor and close the door. And Laika went into space.

Thanks to the skills of the project managers, therefore, the first flight was not delayed. But since there was no gravity in space, it was not possible to determine how a cosmonaut's physique would react.

Gagarin's journey could not be entrusted entirely to an autopilot, so he was given a special code to activate manual control of the spacecraft.

Many things were still unknown: Gagarin could not have known that, during the passage of a vehicle through the dense layers of the atmosphere, there is friction in the heat-resistant coating that appears to create a flame. Gagarin himself, in fact, on seeing the flames, had uttered:

I'm burning, goodbye comrades,

but this was just the normal reaction of the vehicle as it passed through the atmosphere.

Two days before departure, Yuri Gagarin had decided to write a farewell letter to his wife, in case things did not go well. After all, the chances of something going wrong were high:

- The Vostok spacecraft models did not have soft-landing thrusters to ensure a safe descent, and.
- Engineers feared that the hatch would melt in the high temperatures of the dense layers of the atmosphere.

Fortunately, however, all went well in 1961, and the letter written by Gagarin for his wife Valentina was not delivered to her until March 27, 1968, following the plane crash that killed the first man in space. How did this come about?

It all started when, on a day when Valya was in the hospital, Gagarin and his instructor, Vladimir Serjogin, took off in their MiG 15.

What happened next is still a mystery to this day. Gagarin and Serjogin's plane crashed to the ground. Of the two pilots on board, only scraps remained.

2.8 Death

Death would be sweeter if my sight had thy face as my last horizon, and if it had, I would be born a thousand times and die a thousand times more. (William Shakespeare)

When my body is ashes, my name will be legend. (Jim Morrison)

On March 30, 1968, the Soviet Union gave its Hero a state funeral. The Soviet authorities hastened to say that Yuri Gagarin had lost his life in a trivial plane crash. The fiercest and most ruthless detractors suggested that the pilots were drunk. Examinations of the remains of Gagarin's body disproved this.

Leonov, the cosmonaut's close friend, wanted answers. Together with other colleagues, he began his own investigation, from which emerged a hypothesis that was never substantiated: a supersonic Sukoi aircraft, a new concept for that time, had interfered with the flight of Gagarin's old MiG.

In particular, the sound wave produced through breaking of the sound barrier had caused the two pilots to lose control of the aircraft.

The Sukoi, according to what Leonov wanted to prove, should have flown at 10,000 m, and not at 1,000, as seemed to have happened.

In 1986, a new investigation was launched, thanks to Leonov's tireless work. It was discovered that reports had been altered, texts had been changed and modified, and mysterious hands had signed bogus declarations. In short, all the cards on the table had been changed. A mixture of hypocrisy, secrecy and corruption had done the rest.

But now, let us review the chronology of Gagarin's death, as accounted Kamanin's diaries.

After his famous flight, Gagarin managed to get himself appointed, in 1967, as substitute cosmonaut of the controversial Soyuz 1 spacecraft, the "flying coffin," the one that was harshly criticised due to its obvious construction errors, and in which cosmonaut Vladimir Komarov died in dramatic circumstances. From that point on, Gagarin returned to flying the MiG aircraft, as he had done before joining the space project.

After Komarov's death, the Soviet government thought it best for Gagarin to opt for a safer profession. The first man in orbit was too important a symbol for the Soviet Union.

Jurij Alekseevich was then head of the Soviet Cosmonaut Detachment and deputy head of the Cosmonaut Training Centre. He opposed this decision. Kamanin felt the same way: he was afraid that the cosmonaut officers were not preparing properly for flight and were not good enough pilots.

To keep them in training, a squadron had been set up in 1962 under the command of test pilot Vladimir Serjogin, yet another Hero of the Soviet Union.

But Kamanin and the cosmonauts felt that this was not enough.

In 1967, a training regiment was created based on the Serjogin Squadron.

The problems of detaching the cosmonauts from the ship had to be solved: there would be training flights for parachute jumping, and, in addition, the problems of weightlessness (0-G) had to be addressed.

Here is an excerpt from General Kamanin's diary:

February 8, 1968. Gagarin and Titov said they were preparing to pass the exam at Zhukovsky Academy on February 19. This year Gagarin and Titov would be flying a lot: Gagarin was reinstated as a first-class pilot and would be engaged in space flight and training at the GCTC - Gagarin Space Training Center - while Titov would complete his course as a test pilot and candidate for the Mikojan orbital aircraft - that is, the spaceplane designed by the manufacturer Mikojan.

On February 27, 1968, Gagarin and Titov graduated with distinction from Zhukovsky Academy. Gagarin then took up flying, as he had long dreamed of doing. His training flight took place in the MiG-15UTI aircraft.

By the end of March 1968, Gagarin had almost completed his training.

We read again from Nikolai Kamanin's diaries:

26 March 1968. After meeting with the State Commission, General Kuznetsov[3] told me that tomorrow Gagarin is scheduled to be certified for independent flight on the MiG-15. Kuznetsov asked me to personally check the preparations of the MiG-15 on which Gagarin will make the independent flight. But I forbade Gagarin's flight with Kuznetsov, telling Kuznetsov himself that he had long since lost his ability as an instructor pilot. I allowed the regimental commander Serjogin to check Gagarin's flying ability on the day of tomorrow, and General Kuznetsov ordered [him] to check Gagarin's flight organization, to analyze and report on the air situation and weather conditions.

---

[3] Head of the Cosmonaut Training Center from 1963 to 1972.

The language is contrived but clear.

So, on March 27, Gagarin was scheduled to fly. As reported in his diary, Kamanin was in Moscow and could not travel to Chkalovsky to congratulate the pilot on the completion of the programme. Kamanin looked upon Gagarin as a son, but duty came first.

At 10:18 a.m., Gagarin and Serjogin took off in the MiG-15UTI.

The weather was not the best, but the two were very good pilots, exceptional, in fact: for them, what they had to do was an everyday occurrence.

At 10:32 a.m., Gagarin said he was finished and asked permission to return to base.

Then, nothing.

The plane disappeared.

At 10:50 a.m., Kamanin was informed:

With Gagarin's plane, communication has broken down. In ten minutes, the fuel may run out.

Something serious had happened.

But Kamanin thought the matter could be resolved, immediately calling for a rescue plan.

The Air Force leaders converged on Chkalovsky Airfield, where the two had left from.

Planes and helicopters carried out a thorough search.

Again, from General Kamanin's diary:

At 2:50 p.m., the commander of the Mi-4 helicopter, Major Zamychkin, reported: "I found fragments of Gagarin's plane sixty-four kilometres from the Chkalovsky airfield and three kilometres from the village of Novoselovo." "The plane crashed in a very dense forest in the vicinity of the town of Kirzach. At the moment of impact on the ground, the speed was seven hundred to eight hundred kilometres per hour. The engine and forward cabin went six or seven metres underground. The wings, tail, tanks and cockpits were fragmented over a two hundred metre per cent strip. Many tiny pieces of the plane, the pilots' clothes and parachutes were found in the treetops. After a while, a fragment of an upper jaw with one gold and one steel tooth was found. The doctors said it was Serjogin's. There was no sign of Gagarin's death, but hopes of his salvation fell."

From Moscow, the government wanted certainty: what had happened to Gagarin?

Experts on the ground had no doubt.

He was dead.

But Kamanin wanted an answer that was absolutely certain.

On March 28, he found it.

Again from his diary:

> Around eight o'clock in the morning, General Kutakhov and I noticed a fragment of material on one of the birch trees, at a height of ten to twelve metres. It was part of Gagarin's jacket. In the jacket pocket we found a breakfast coupon with the name of Yuri Alekseevich Gagarin on it. The doubts were gone: Gagarin was dead.

He was only thirty-four years old.

The army found and collected the remains of the bodies, which were cremated the same evening.

On March 30, 1968, the crowds that gathered were immense. The funeral of the hero Yuri Gagarin was being celebrated. The urns containing the ashes of Gagarin and Serjogin were buried in niches in the Kremlin walls.

Gagarin's death affected the whole world.

One wondered how this could have possibly happened.

Documents concerning the investigation into the deaths of Gagarin and Serjogin were classified for years in the USSR.

So many conspiracy theories arose. According to some, Gagarin did not like Soviet power, so he was "deliberately eliminated." According to others, Gagarin was alive, but mentally unstable, and was sequestered in a psychiatric hospital, where he had died.

But compared to the other cosmonauts, Gagarin appeared to be far more balanced. He did not drink as much as the others, and was, in fact, a model cosmonaut.

The question remained: what went wrong?

Fingerprints were taken from the dashboard and Gagarin's wristwatch; the latter indicated that the accident had occurred between fifty and seventy seconds after the final communication.

The forensic experts tested the pilots' blood and declared that it was "clean."

It seems that, until the last second, the pilots were trying to save the plane, but nothing was wrong with it. It was working perfectly.

In 2011, more papers were published: they posited another possible cause of the disaster.

Perhaps a very fast manoeuvre had become necessary to escape from a weather balloon. By an incredible coincidence, that balloon had found itself on a collision course with Gagarin and Serjogin's plane. The lightning-fast manoeuvre had sent the plane into a spin caused by making three, four or five turns. This was the assumption. At the same time, the two pilots had to deal with a 10–11 G load, but the plane kept flying. According to the experts, Gagarin and Serjogin could do nothing, being only two or three hundred metres above the ground before crashing.

Aleksej Leonov, the first Russian to do a spacewalk, revealed a different version in a 2013 interview. His version is the most widely accepted of the various theories about the cause of Gagarin's crash:

> On that day, a training flight was conducted not only by us cosmonauts, but also by an Su-15 aircraft on behalf of the Air Force Flight Test Institute. There was one rule, however: our flights were at a maximum altitude of ten thousand metres, the others were much higher. But the Su-15 violated the flight regime, descending to an altitude of about four hundred and fifty metres. At that moment, local farmers spotted it. After flying for a while at this altitude, the Su-15 left for its training flight, where it should have been. There were thick clouds, with Jurij Gagarin flying at four thousand two hundred metres. The Su-15 flew at supersonic speed and the wake created turbulence, pulling Gagarin's plane into a spiral screw, and he descended from four thousand two hundred meters to zero altitude (i.e., colliding with the ground) in fifty-five seconds.

Leonov said it was all amply demonstrated. Here are his words:

> Forty-five years after the disaster, I asked President Vladimir Putin to open a secret document. The president allowed it. Now the fact, already claimed, that the cause of Gagarin's death was an unauthorised flight is confirmed.

In an interview with RIA Novosti, Leonov revealed that he knew the name of the Su-15 pilot:

> I was given the opportunity to announce the real cause of the accident, but I did not mention the name of the pilot, who is now 80 years old and in poor health. I gave my word not to name him.

But there is another hypothesis that refutes Leonov's. It is that of Arsenij Mironov, one of the oldest specialists at the Gromov Flight Research Centre. Mironov celebrated his 100th birthday in December 2017 and died in 2019. A few months earlier, he had published an open letter in the tabloid *Komsomolskaja Pravda* addressed to Aleksej Leonov, a letter that dealt with evidence that disproved his version:

You said that the Su-15 was flown by a pilot from the M. M. Gromov Flight Research Centre. Apparently, for human considerations, you have not named the pilot so far. But the blame for the death of the planet's first cosmonaut is now being placed on the LII research institute! I know that on 27 March 1968 it was decided that two Su-15 aircraft would fly for the LII. The first was piloted by the experienced test pilot Arkadj Pavlovich Bogorodskij. It flew to test a new engine at an altitude of eighteen thousand metres. It took off at 9.45am and landed at 10.24am. Due to the high altitude of the flight and the possibility of depressurisation of the cockpit in the event of engine shutdown, the pilot wore a spacesuit, which impeded his movements and with which he could not eject. The flight of the second aircraft was carried out by Hero of the Soviet Union test pilot Aleksandr Aleksandrovich Sherbakov at an altitude of fourteen thousand metres, taking off at 11:20 a.m., after the MiG-15UTI had crashed. Time analysis shows that Bogorodsky landed while Gagarin was still gaining altitude. Sherbakov flew long after the MiG-15UTI crash, so the two Su-15s would not in principle have prevented the MiG-15UTI from flying.

Mironov thought that Leonov's version had been debunked precisely on the basis of the information given by LII Gromov. But Leonov continued to present his version in various articles and books. Mironov believed that the version of the ball-buster was a palliative to silence everything. For him, the reason for the disaster was a lack of proper flight planning. This would have resulted in a risky approach, not with the Su-15, but with another MiG-15UTI that had taken off from Chkalovsky Airfield.

In 1968, the State Security Committee wanted to know whether Gagarin had died as the victim of a planned failure. The investigation found a few minor infractions, but absolutely no sabotage. However, it is still widely believed that Gagarin survived. It is thought that he was fed up with popularity and wanted to live in peace. So, he faked his own death. After that, he went to a small village near Orenburg, and ended his days under a false name, hunting and fishing. This is a very common legend about the disappearance of heroes.

The figure of Gagarin was used extensively by Soviet propaganda, not only to assert the supremacy of the USSR in the space race, but also in favour of state atheism. In fact, Gagarin was credited with the phrase:

I see no God up here.

But the attribution of the phrase is controversial; there is no record of it, so it could be the result of Soviet anti-religious propaganda.

Gagarin's friend and colleague Colonel Valentin Vasil'evič Petrov, a lecturer at the military aviation academy named after Gagarin, gave an interview to the Interfax news agency in 2006. In this interview, Petrov stated that Yuri was baptised in the Orthodox Church and was a believer, and he had several direct testimonies of his sensitivity to the divine. The colonel was convinced that it was, in fact, Nikita Khrushchev who coined the famous phrase about the absence of God in space. Petrov recalled the statement of the First Secretary of the PCUS:

Why are you clinging to God? Gagarin flew in space and did not see God.

Leonov himself, interviewed by *Foma* magazine (No. 4, 2006), reported that Gagarin wanted to baptise his daughter Elena shortly before the flight. He also stated that the Gagarin family celebrated Christmas and Easter every year, as well as having religious icons and images in their home.

Back to Gagarin's death. It is March 27, 1968, and seven years have passed since the feat that made him a hero. The man is flying aboard a small MiG-15UTI fighter jet and suddenly crashes to the ground. A freak accident, mysterious. A sudden dive and then the crash.

What could have caused the death of an experienced aviator, as well as a cosmonaut, in a flight that was supposed to be a piece of cake? There was talk of alcohol, conspiracy, a faked death for the purpose of being reborn to a new life, murder.

In these forty years, every possible explanation has been heard.

Questions have even been asked about the Soviet leader Leonid Brezhnev, who, envious of Gagarin's success, allegedly had the plane sabotaged. Over the years, a thousand hypotheses have fostered the birth of the legend. This is yet another mystery hidden in the archives of the Soviet Union.

Even former Air Force Colonel Igor Kuznetsov, after taking part in the initial investigations and working for nine years to solve this mystery, told the British newspaper *Daily Telegraph* about the possible cause of the sudden "crash."

According to him, on that day in March 1968, Gagarin and his co-pilot Vladimir Serjogin were conducting a routine flight, when the cosmonaut suddenly noticed that an air vent in the cockpit had been left open. The cabin was not properly pressurised and the plane was three thousand metres above the ground. Gagarin was startled. He panicked, thinking that the only way to save his life was to dive to a safer altitude. Descend at over a hundred and forty-five metres per second. Fast. So as not to die. In those days, pilots didn't know that such a sudden and rapid descent could cause enormous damage. As a result, the two lost consciousness and crashed in the nearby Kiržač forest.

Colonel Kuznetsov, together with his staff, used the latest investigative techniques and consulted hundreds of documents to determine what caused the crash. Kuznetsov is still firmly convinced of his conclusions and is calling for the case to be reopened, a request that was denied by Russian President Vladimir Putin in 2007. Why?

After more than 50 years, Gagarin's family have the right to know how one of Russia's most famous national heroes died.

But the conclusion is still shrouded in darkness.

3

Female Cosmonauts Without Flights (Promises not Kept)

Abstract The reasons why the USSR was the first to send a woman into orbit are analysed. The women selected for the flight are discussed (there were five of them, and one of them, who has since become a historian, has much to say). Then, the image of one of the most deserving is sketched and the question is asked as to why she was rejected. Finally, a portrait of the only one who actually flew (apart from Tereskhova, who would be the first woman in space) is given and the relationship between the women and their male colleagues is examined.

3.1 Introduction

Astronauts are ordinary people doing extraordinary things.

(Sandra Bullock)

If you haven't encountered difficulties in your life, go out and buy them. Only by facing them will you be able to call yourself a man.

(Japanese Proverb)

At the beginning of the 1960s, few women in the world were able to work in science. But the Soviet Union was an exception to most countries on earth. The Marxist doctrine of the regime had no gender bias. At least, in science.

M. R. Menzio, *The Secrets of Soviet Cosmonauts*,
https://doi.org/10.1007/978-3-031-09652-5_3

That is to say, women and men were considered perfectly equivalent. It is this kind of mentality that allowed Valentina Tereskhova to become the first woman in space.

A few figures: in 1964, in the USSR, forty percent of engineering graduates were female.

In the same year, in the USA, they accounted for less than five per cent. Score one for the USSR.

In the mid-1980s, fifty-eight percent of engineers in the USSR were women.

In the US, it remained below twenty percent. Two to zero for the USSR.

The regular Soviet school curriculum favoured the study of mathematics and the hard sciences. So, women and men tried to establish themselves in these fields.

Tereskhova's feat proved that a properly trained woman could work in space just as well as her male colleagues. That flight had an enormous propagandistic, political and social impact during the Cold War. In the USSR and abroad, the cosmonaut was regarded as a symbol of Soviet technological and social achievements, as well as a role model and an example of female emancipation. But none of Tereskhova's companions chosen in 1962 were ultimately able to go into space, and the group was disbanded in 1969.

At the same time, the "Mercury 13" did not have any better luck in the United States. These were a group of thirteen women who were chosen as finalists in 1960 as part of a private research project. They had passed the same selection tests as the male Mercury astronauts. Among them was the legendary aviatrix Geraldyne Cobb, whom Kamanin met during his travels.

Geraldyne (sometimes known as "Jerry") had flown for the first time at the age of 12 in a biplane piloted by her father. From then on, flying became her reason for living. She herself said:

I feel that life is a spiritual adventure, and I want to live mine in the sky.

At the age of twenty-nine, Jerry Cobb was a world-class aviator.

She then underwent medical and psychiatric examinations, sensory isolation tests and flight simulations for Project Mercury.

She passed all the tests with flying colours.

She quickly became a star, being interviewed by all the newspapers.

But that's where the macho vision of the American man on the street came in. Everyone wanted to know why a pretty girl like her wouldn't want to spend her time just thinking about clothes and boyfriends. Here's proof of that from an interview:

Interviewer: "Do you think you can compete with men?"

I don't think I have to compete with men; I think both men and women should fly together in space.

Interviewer: "A nice girl like you should be thinking about marriage, don't you think?".

Right now, I'm interested in this space flight and nothing else.

Over the next few months, another twenty-five women were selected for the project. Among them, twelve passed the medical tests required for space flight, in some cases, with better results than their male counterparts. These girls formed the Mercury 13 team.

However, the desire of many women to become astronauts was still being ignored by NASA. This was clearly discrimination.

The Mercury 13 women stepped forward and demanded an official debate on their project.

A committee of the US Congress was convened.

But, after carefully studying the case, the committee concluded that:

NASA's selection programme has been conducted in a reasonable and expedient manner, and it would be good to continue to maintain the highest possible standards. At some point in the future, consideration may be given to initiating a research program to determine the advantages of employing women astronauts.

Off the record, one congressman said:

It is our social order that dictates the differences between men and women. The space programme merely follows the gender differences imposed by time and history. I don't think American women want to do everything that Russian women are **forced** to do.

There were also wry comments such as:

Male astronauts are absolutely in favour of employing women as astronauts. And we are reserving fifty kilograms of payload for recreational equipment.

The result: neither Geraldyne nor the others could convince NASA to accept them as astronauts to fly in space.

Worldwide, women would have to wait until 1982. As we shall see later, only nineteen years after Tereskhova's flight, it was again a Soviet woman,

Svetlana Savitskaya, who flew a mission on the Soyuz T-7 spacecraft to the Salyut 7 space station. And on a subsequent mission, Savitskaya was also the first woman to perform a spacewalk.

The following year, on June 18, 1983, it was finally the United States' turn: NASA astronaut Sally Ride became the first American (and the first gay woman) to fly into space on the Shuttle STS-7 mission.

One might recall a beautiful phrase uttered by the scientist Chien-Shiung Wu (the woman who discovered the violation of the parity principle in weak interactions):

> It would be interesting to ask elementary particles whether they would enjoy being studied by a man or a woman more![1]

But how did the idea of launching a woman into space come about in the first place?

Historian Margaret Weitekamp, in "Right Stuff, Wrong Sex", writes:

> Before any person had flown in space, some researchers had wondered whether women were better suited than men for space flight. Scientists knew that women, smaller beings on average, required less food, water, and oxygen, which was an advantage when flying both a person and his or her supplies in a confined space vehicle.

The Mercury 13 scientists[2] found that, in isolation tests, women performed better than men. They often had stronger cardiovascular health.

This project was conducted by NASA specialists, but was never on the agency's official agenda. It was a privately funded initiative. And, at that time, it failed to change gender policies in that area. In other words, male chauvinism continued to prevail in the space sector.

---

[1] Apparently, conservation of parity is lacking in physics, but also in interactions between people: Chien-Shiung Wu did not win the Nobel Prize for her revolutionary discovery, but was only rewarded later with the Wolf Prize, the National Medal of Science and several honorary degrees. She even got her own postage stamp. But the Nobel Prize for physics went to the physicists Tsung Dao Lee and Chen Ning Yang, who formulated the theory, leaving the scientist who proved it in the shadows.

[2] Mercury 13 (or Fellow Lady Astronaut Trainees) was a space programme carried out in the United States of America in the early 1960s on the initiative of researcher William Randolph Lovelace II, who had arranged medical tests for the selection of astronauts for NASA. It provided for the selection and training of women as pilots for the first US astronaut missions.

3.2   John Glenn's Barbecue

Without pots, without pans, without water, without oil, with only the aid of fire and a piece of iron on which to lay the raw meat, the act of cooking regains the primordial sense it must have had as soon as Prometheus gave fire to men. (Massimo Montanari).

Nikolai Kamanin travelled often.

Abroad, he tirelessly supported his nation's space efforts. From April 1961 to January 1963, together with Gagarin and Titov, he visited more than thirty countries, an itinerary that included the United States. There, Kamanin and Titov met with President John F. Kennedy, and dined at the home of John Glenn, who had been the first American to orbit the Earth in the Mercury 6 capsule.

Khrushchev did not come up with the idea of launching a woman into space all by himself. This is a common misconception. It is true that Khrushchev was prime minister at the time, and was very attentive to the space programme and its propagandistic aspects. It is true that the Soviet Khrushchev liked the idea of launching a female cosmonaut very much. But it was not he who came up with the idea of including women in the space programme.

The idea of launching a woman came from Nikolai Kamanin, who, in March 1962, had his five female cosmonauts to train: Tatjana Kuznetsova, Valentina Ponomareva, Irina Solovjova, Valentina Tereskhova and Zhanna Yorkina.

This is where John Glenn comes in. It was in May 1962 that the Soviet delegation, with Kamanin and Titov, met with Glenn.

The two Soviets are invited to a barbecue in the astronaut's garden. Beer in hand, the guests talk about this and that, the weather, politics and the conquest of space. Kamanin is nervous, he doesn't feel well, it's hot, and, above all, there's no trace of vodka, his beloved vodka!

Glenn cooks the steaks and watches the Russians roast the sausages.

Meanwhile, he talks. The idea that seems to be closest to his heart is that of flying a woman in Mercury capsules (although, as it turned out, the idea was only close to his heart, as opposed to being close to fruition).

We can imagine how the meeting went:

Kamanin: It's hot in here! I'm used to Russian winters.

Glenn: Sure. You have Siberia, land of minerals... (*puts a steak on the grill*) and gulags.

Kamanin: You (*puts a sausage on the grill*) closed your embassy in Cuba not too long ago, didn't you?

Glenn: (laughs) Did you have the pleasure of hearing President Kennedy's speech?

Kamanin: Did you have the pleasure of seeing the May Day parade with Khrushchev in Red Square?

Glenn: (laughs again) You guys are pioneers in space. (*turns over steak*) Just like us.

Kamanin: Are you going to go ahead with the space programme, or not, given the many... delays? (*turns over sausage*)

Glenn: By the way, any news about Laika? (*eats a piece of steak*)

Kamanin: Anyway, we were the ones who sent the first man into space. (*eats a piece of sausage*)

Glenn: (*tears off a piece of steak*) But there are women too, or has communism forgotten that?

And thus, Glenn, an enthusiastic supporter of the "Mercury 13"—female pilots lobbying to be trained as Mercury astronauts—became the person responsible for Valentina's launch.

In fact, Kamanin didn't understand that being a supporter only meant that Glenn liked the idea. He misinterpreted Glenn's words to mean that the first American woman astronaut would be part of a three-orbit Mercury flight by the end of 1962.

However, NASA was not yet planning to launch a woman into space. The agency made this position clear in response to a letter sent by Linda Halpern, a primary school student. The girl asked President Kennedy how to become an astronaut.

We have no plans to employ women in space flight,

NASA replied.

But Kamanin returned to the USSR convinced that the Americans were intent on launching a woman into orbit, and could do so in a very short time. The Soviets had to hurry to keep the US from stealing a record from them.

With a (metaphorical) knife between his teeth, Kamanin announced to the Russian government that "*the Americans will beat us.*" So, within weeks of his return, he was given the go-ahead. In a short time, he was to be in charge of facilitating the first flight of a Soviet woman—and not just one woman.

By May 1962, when the Russian delegation visited the United States, the first Soviet trainees had already been accepted into the space team in Star City (a town about 30 kms from Moscow, where the cosmonauts live and train).

3.3   Selecting the First Female Cosmonauts

Moscow officially approved the idea of launching a cosmonaut into space. As a result, more than eight hundred women applied for the positions. After the first screening, fifty-eight remained to be formally considered. In the semi-finals, only twenty-three candidates were chosen for advanced medical screening, i.e., the list of checks to be carried out in Moscow.

The cosmonaut candidate could be a university graduate, but this was not mandatory. The focus was on the skills needed to be an astronaut. But this led to a problem. The male candidates had been selected among Air Force test pilots. This career path was not then possible for Soviet women.

There were some women who had similar qualifications. After the war, it was not difficult to find female aviators who had not only served during the Second World War, but had also participated in air battles. However, all of these female veterans were well into their thirties.

There were not enough women. What to do?

The Soviet government decided to go a different way. It sent out scouts to look for cosmonauts in local skydiving clubs. This sport had been very successful throughout Russia since the 1930s. At that time, the government decided to encourage the sport among all young people to prepare them for the next great war.

So, parachuting was considered a relevant qualification. Why? Because of the way the first Soviet cosmonauts left their capsule. In fact, as we have seen, the first spacecraft models required them to use a parachute for landing. When the women's unit was put together in Star City, Soviet engineers had not yet worked out a safer landing strategy.

The finalists for the women's space team were divided into two groups and health screening tests began in January 1962. The women underwent medical examinations in the same hospital as pilot Alexey Maresyev. Maresyev had lost both legs in combat, but had tried to prove to a group of doctors that he could still fly. They were amused. According to legend, Maresyev demonstrated his physical prowess by performing *Gopak*, a Ukrainian Cossack dance.

Following the same protocol used for male candidates, the women were subjected to multiple health and psychological examinations. The various doctors took x-rays of their bodies, studied their brain functions, did blood tests and cardiovascular checks. The women also underwent centrifuge training, in which a machine rotated rapidly to apply powerful centrifugal forces to the cosmonaut. The scientists used this test to determine how the subjects would handle acceleration in zero gravity.

In the end, only five women remained.

Who were they? ZHANNA YORKINA, VALENTINA PONOMAREVA, IRINA SOLOVYOVA, TATIANA KUZNETSOVA, and VALENTINA TERESKHOVA. The last to be rejected before the final five were settled upon was named TATIANA MOROZYCHEVA.

We will now discuss the first four candidates. We will then take a closer look at Morozycheva, as well as Marina Popovich and Svetlana Savitskaya, two other female stars of the Soviet space programme.

3.4   Zhanna Yorkina, the Foodie

The paratrooper Zhanna Yorkina was twenty-five years old, had been taught in a country school and was chubby. Not fat, mind you: just soft. But she was a unique and qualified candidate. In addition to being a paratrooper, she spoke two foreign languages, German and French. But these skills did not help her during the centrifugation test. Zhanna recalled that:

> I weighed 132 pounds (60 kg), but because of the g-acceleration, I felt an extra 600 kilos of pressure while in the capsule. [...] I didn't like this. If you relax your abdomen, you lose consciousness, which has often happened to men as well. [...] We were holding a remote control during the test. The doctors said, with a certain malice: "If we hear you holding the remote control, it means you are conscious. If you let go of the remote, you're unconscious. Then they take you outside and you are excluded".

Yorkina was considered one of the least capable of the five cosmonauts. Kamanin often complained about her.

"Too fond of chocolate and cakes".

All things aside, how exciting it must have been for her, at just 25 years of age, to have been in contention for the very first space flight by a woman. Within a matter of a few weeks, in early 1962, Yorkina was plucked from the obscurity of a provincial flying club, where she was a competent hobby-parachutist, and taken to the embryonic 'Star City,' where she was plunged into a military environment and faced the very real possibility of being strapped on top of a rocket within six months. There is no questioning her bravery, or her sense of adventure. History suggests that she was always an outside bet for the first female spaceflight, yet, she was still under sufficient consideration to undertake the final pre-selection tests in April/May 1963, which included a three-day 'launch to landing' simulation, in the Vostok mock-up. It is sad that the circumstances of the time meant that she and her colleagues (except for Tereskhova) spent the next 25 years hidden from public view, and therefore she never really received the recognition, or acclaim, that she probably deserved. That is incredible!

3.5 Valentina Ponomareva, the Refined Free Thinker

At the start of the selection process, Valentina Ponomareva was twenty-eight years old.

She worked in the Department of Applied Mathematics at the Steklov Mathematical Institute, which was part of the Russian Academy of Sciences. The Institute was closely linked to the design office headed by Sergey Korolev.

Smart and educated, she graduated from the Moscow Institute of Aviation. She had chosen a career in mathematics, instead of following her passion for literature, a passion that she had had since her high school days.

But there was also a third dream that she wanted to fulfil: life in the sky, skydiving, space flight.

As a university student, Ponomareva skipped classes to work and fly with a local aviation club. There, she met another amateur pilot who later became her husband and the father of her child.

While dancing with a colleague at a New Year's Eve party, she received an offer she hadn't expected: to try to "fly higher than any pilot." Ponomareva said yes without hesitation, even though she thought it was a joke. Her colleague was insistent, and Ponomareva eventually sent an official application to her new boss, Mstislav Keldysh, who was later promoted to President of the USSR Academy of Sciences.

When they met, Ponomareva was nervous. In her eyes, Keldysh was an exceptional figure, given his extraordinary contribution to the Soviet space industry.

"Why do you like to fly?" Keldysh asked of her.

"I don't know," Ponomareva replied sincerely.

"That's right, we can never know why we like to fly," said Keldysh, and he accepted her application.

In one of her books, *The Feminine Face of the Cosmos*, Ponomareva states:

> *When the DRCD index jumped next to a window, the window would go out, a signal would sound, and all this had to be reported back to Earth. "The DRCD index has started," the cosmonaut would say. I liked the fact that those words were full of hidden meaning and could only be understood by the initiated.*
>
> *Korolev asked me whether I would resent it if I did not get to fly. I said: "Yes, Sergei Pavlovich, I would resent it very much!" He said: "You are right. I would have resented it too." Then, he gave every one of us a look, and said: "It's all right, you'll all fly into space."*
>
> *It was announced that Valentina Tereskhova was appointed the commander, and Solov'eva and Ponomareva were the back-ups. I remember Karpov's explanation: two back-ups, instead of one, as it was with men, "with consideration for the complexity of the female organism."*
>
> *On the first Soyuz ships, there was no onboard computer, and in case of failure of automatic systems, it was impossible to carry out manual docking. This was because the back-up manual guidance system on board did not ensure effective participation of the cosmonaut in the control process. I think that the situation was aggravated by the discrepancy between the terrestrial stereotype of relative movement and the reality of space flight: we got used to relying on our experience of operating airplanes and automobiles, where it is possible to "add gas" to catch up with a moving object. ... Besides, a significant role in guiding an airplane belongs to intuition. ... In order to predict the relative movement of objects, it is necessary to know their orbits. This idea is confirmed by the fact that, when an onboard computer was installed, in case of failure of automatic systems, dockings were performed manually with success.*
>
> *The requirements [for a cosmonaut] are very strict. Readiness to take risks, sense of responsibility, ability to carry out complex tasks in harsh conditions, high dependability of the operator's work, advanced intellectual abilities, and physical fortitude. ... But the cosmonaut must also possess such qualities as the ability to break the rules. ... The ability to act in extraordinary situations is a special quality. In an extreme situation, the very life of the cosmonaut depends on these qualities. Korolev captured his vision of professional qualifications for the cosmonaut: for cosmonauts, one must not select the disciplined, but rather the intelligent.*

Ponomareva herself went on to clarify another aspect of the space programme: the reason why the Russians lagged behind the Americans in the second half of the 1960s.

When the first cosmonauts and astronauts were in orbit in their spacecrafts, be it Vostok or Mercury, the premises and lines of thought that led to their tasks were the same in Russia as they were in the US, and the evolution of guidance systems was also the same.

After all, no one knew whether a human being would survive a space flight, and if so, how he would do so. Moreover, there were precedents in unmanned space vehicles for automatic guidance systems. Finally, the most important question, the safety of the pilot, was linked to that of the control of the vehicle itself.

In the beginning, the Soviets and the Americans each focused their line of thinking on both guidance designs and the importance of automatic guidance. Thus, all control procedures were automated, and only in emergencies was there a switch to manual driving. The pilot was excluded from control of the spacecraft.

But the technical feasibility for this similar approach was different in the two countries: this was because, while the manned Vostoks weighed 4.5 tonnes, the Mercurys weighed only 1.3–1.8 tonnes. A decisive difference in on-board systems, and a difference in the designers' approach to the tasks of humans in space vehicles.

Flight reliability is ensured if the safety of the structures on board the vehicle is increased. To do this, it has been traditional practice to at least duplicate (make redundant) critical system components. One is reminded of the philosopher of science Paul K. Feyerabend and his book *Conquest of Abundance*, the story of the clash between abstraction and the richness of Being.

On the Soviet side, the weight of the Vostok made it possible to duplicate almost all of the vital systems, except for the braking engine unit. These systems were installed on board, so the spacecraft was quite safe and fully automated. The cosmonaut only had to keep an eye on the operation of the vehicle itself.

On the other side of the world, the Mercury weighed little, and thus limited the ability to duplicate systems, which corresponded to a low level of safety. To increase reliability, on-board systems were copied with manual controls. Thus, the US astronauts had a fairly broad area of control, covering on-board systems, changing flight plans, and even knowing how to approach or leave orbit in the most dangerous cases. Their choices were not dictated by ground control, but rather by the information of the moment, derived from the flight path received on board. In the early flights, guidance systems often failed, and the astronauts had to control a lot of data by not only monitoring it, but actually changing it. Certainly, the people aboard the first spacecraft had opposing guidance systems on opposite sides of the world.

There were also other reasons for this. Behind the two different choices was the history of the basic technology for manned spacecraft. In the US, this technology followed the rules of aviation, i.e., extreme reliance on the pilot.

In the USSR, the technology was based on rocketry. The rocket designers had no people in the air, so they understood perfectly well the idea of automatic control.

What's more, the Vostok and Mercury launches forged the respective future models of the USSR and US space programmes with people on board. On Gemini, guidance was mainly manual, for example, in situations such as approach and docking. In this way, the operation of a system with defective parts was made safe: if the technology failed, the skill of the pilot intervened. On the Vostok, it was the other way around, although, in the beginning, the importance of automatic control was justified. But the further increase of automation and the distrust in the pilot continued with the Soyuz, even when this system had failed, on both manned and unmanned flights: examples were Vostok 2 and, especially, Komarov's flight.

Why was this so? Initially, Soyuz was to be part of a set of lunar rockets to be mounted in orbit. To do this, it needed five docking stations, four of which were made up of unmanned vehicles. Automatic control for approach and docking was an essential element. Then, there is the recurring idea in technology: if a certain approach is successful in a certain sector, then that approach will be used going forward in developing that sector. This is how manned spaceflight developed in the USSR and the USA.

But the main reason for the development of automation was the Soviet technological orientation of those years, since there was an objective opportunity for automatic control established by the technology of the various systems. The focus was on the machine: it was thought that the safety of manned spacecraft could only be guaranteed by the safety of technology. Automation therefore replaced the function of the pilots. But one factor was not taken into account: neither the "aircraft-pilot" nor the "spacecraft-pilot" complexes were isolated, since the human and technical part on the ground also contributed to flight control. However, their reliability was never analysed.

Subsequent Soviet examples of test docking showed the errors of this line of thinking. Criticality in the automatic systems, rendezvous and incomplete docking in tests: none of this convinced the designers of the problems inherent in automatic control; rather, they flew headlong into eliminating the pilot from control. As the philosopher Ralph Waldo Emerson said:

A foolish consistency is the hobgoblin of little minds.

This was the reason why the Soviets lagged behind the US.

All we need to do is make a comparison between Soyuz and Gemini, designed in the same years. Such situations have shown that you cannot create a completely reliable automatic structure: a technical failure will occur, and immediately afterwards, a human person will have to take action. That person must then be able to take over from the faulty system. But if they cannot do this, they cannot control anything. If they could, they would have to be trained in manual control, either by training on the ground or during the flight. If, because of automated flight, the pilot does not have all these skills but has only observer functions, they cannot do the right thing in an emergency.

The obstinacy to find a completely safe automated system leads to multiple redundancies in control, high costs and increased weight of the spacecraft. All of this highlights the error of the automatic approach in spacecraft with cosmonauts on board. After much stubbornness, the Soviets admitted that, yes, the pilot did play an essential role in the functionality and reliability of spaceflight… and eventually switched to semi-automatic structures, but it was too late to close the gap with the US in the race to reach the Moon.

In the US, the Gemini and Apollo programmes had many failures and emergencies, but they were still able to complete their flights thanks to the school of thought of semi-automatic control methods, in which the function of the pilots was essential. While Gemini had a semi-automatic type of guidance, Apollo was built so that the pilot would be able to carry out all of the manoeuvres necessary to return home from anywhere within the Moon's orbit, regardless of the data coming from the Earth.

Eventually, the two antithetical visions met. Semi-automatic designs were arrived at by both the Soviets, who came from fully automatic systems, and the Americans, who came from manual systems.

In 2002, Valentina Ponomareva gave a famous interview, conducted and translated from Russian by Slava Gerovitch.[3]

Here is a summary.

Ponomareva, who has become a historian, tells how the first group of female cosmonauts was formed. What strikes her most about the memoirs of the government and rocket designers is the exaggeration. It's amazing how people who were trained and worked at the highest level could think:

> …that cosmonautics would develop at an exceptional rate, spaceflight would become regular and routine, and almost as many spacecraft would be built as aeroplanes.

---

[3] From Slava Gerovitch, Voices of the Soviet Space Program: Cosmonauts, Soldiers, and Engineers Who Took the USSR into Space, Palgrave Macmillan, 2014.

They were all aware of the difficulties and costs of such a programme, yet they were grossly deluded.

> At the end of 1961 Korolev sent a letter to Kamanin, in which he wrote that in the near future 60 cosmonauts would be needed, and among them should be 5 women. Only 20 cosmonauts were then trained. Such were the assumptions on which our space programme was developed.

As always, it all started with the unrelenting struggle between the USSR and the US. You couldn't come second. And manned launches were what impressed the people the most. In the US, says Ponomareva, the "Mercury girls," including first-class pilots, had asked the vice-president to include them in the space programme. Their request was not granted, but the whole world heard about it. In Russia, the problem with historiography is the secrecy and cover-up imposed on many episodes by the regime at the time. Anyway, they wanted to spend about six months preparing some women for the flight. But there was a difficulty. Neither the suits nor the rockets were ready by the appointed date. So, the training of the women was postponed. Selection took place among members of aviation clubs, but, in truth, they were looking for women who were good at parachute jumps, because the cosmonauts had to land in parachutes. It's a difficult technique, and you can't train someone who's almost completely unfamiliar with it in a short time. Ponomareva says that the final group of five women included four jumpers of different levels and herself. But she was a pilot, and as a parachutist, she had made only eight jumps, a number that pales in comparison to the eight hundred jumps of a master parachutist like Irina Solov'eva.

> Kamanin wanted to examine the records of some 200 female aviation sport candidates and asked the Central Committee of the Voluntary Association for the Advancement of the Army, Air Force and Navy for help. They could only find 58 women, and he was rather disappointed. He wrote in his diary that the Central Committee of the Association had done a bad job and that 58 candidates were not enough. At that time, everything related to cosmonautics had to be on a giant scale. From those 58 candidates, the five finalists were selected.

The women trained with the same programme as the males, which mainly involved adapting to the absence of gravity and loads of various Gs. When they arrived at a military training centre, many did not understand the regulations or accept the discipline. When Kamanin asked them all if they wanted to become regular Air Force officers, the women went into a tizzy. They were advised to say yes by their male colleagues, who, at first, were against the women's group, but later came to think that it was a matter of prestige to be first in space with a Soviet woman.

> After Tereskhova's flight, the Centre commanders wanted to get rid of us. But the fact that we were regular officers was an obstacle.

However, the most important type of preparation concerned the resistance of the human body to space flight. In those days, this was unknown, especially concerning the lack of gravity, the only factor that could not be reproduced on the ground. For multiple G-loads, however, the solution was quite simple: use a centrifuge. Initially, the trainers exaggerated: for the males, they would go up to 12 G, for the women, up to 10 G, obviously all done in a gradual way.

> True weightlessness was simulated with flights, first in fighter planes and then in a huge flight laboratory. It lasted 20–40 seconds. Time to notice that a pencil sharpener was floating in front of you.... Then you could see a cosmonaut floating in a spacesuit. But the absence of gravity on the ground is not the same as the absence of gravity in orbit.

The most relevant skill was always the parachute jump. However, for the Vostok spacecraft, there was initially no desire for a manual guidance method. These were the first flights, nobody knew what was going to happen to the crew, so they had to rely on automation. Some doctors were afraid that the humans would freak out! So, the engineers prevented the descent engine from being controlled by the spacecraft. There was a secret code, given in a sealed envelope to the cosmonaut just before the launch to allow for manual control of the descent.

> We knew that one person had given that code to Gagarin, but recent records show that there were four or five such informants.

But it didn't change the engineers' view: they wanted to leave everything to automation. Konstantin Feoktistov, a cosmonaut and engineer, said that a cosmonaut's job was not to waste time driving a spacecraft, but to do research. But if a malfunction occurs, the pilot who cannot control anything is disarmed.

> On the bridge, the sphere in the middle is called "the globe"... it shows the rotation of the Earth and the movement of the ship. If you "reset" the globe,

you will see where the ship will land if you turn on the descent engine at this time. Above the globe is a digital indicator of the number of orbits.... Then four indicators of humidity, temperature and pressure inside the capsule, oxygen and nitrogen pressure in the capsule and pressure in the pneumatic systems of two correction systems. On the right, a set of "windows". In the event of an emergency, a specific message would light up in each 'window', such as 'enough gas for descent only'. For each emergency, there was a corresponding 'window'.

But the instrument panel served only an informational function. The cosmonaut could not do anything active, only turn on the descent engine correction system.

As to the overall question of whether women should be in the space programme in the first place, there was also superstition on Korolev's part: he thought that having women on a launch pad would bring bad luck!

In the end, the women discovered that they had no future as cosmonauts. The most important thing had been done: Tereskhova's flight had yielded results that had been predicted by propaganda, and, in future, only men would fly. Korolev's death put an end to female flights.

But, in 1966, Kamanin wanted to fly them again. He summoned Ponomareva and Solovyova and said that there was talk of them flying over Voskhod for a good 15 days. Once again, an exaggerated project! Until then, the longest flight had been 5 days, with Bykovskii. In 1970, on Soyuz 9, Nikolaev and Sevast'ianov would fly for 18 days, but would also come back barely alive!

They were training when the news of Korolev's death arrived. After that, no more flights for them. And as to the USSR-US competition, the Soviets were starting to fall behind. Komarov's dreadful death on the first Soyuz had, in part, been due to the rush to produce a grand result on the 50th anniversary of the October Revolution. If Korolev had still been alive, he would never have given permission for Komarov's tragic flight.

It was not known whether he would return alive, but he was officially saying goodbye to Africa, Asia and Australia. All was well according to TASS. But the ship was in danger.

3.6 Irina Solovyova, the Skydiving Champion

Before being hired as a cosmonaut, Solovyova was a champion of the Soviet national parachutists, having about eight hundred jumps to her credit. There is not much information about her.

When she was approached to fly in space, she was a 24-year-old engineer from Ural. Some sources say she had two engineering specialties, others point out that she also had a science degree.

> My skydiving instructor and future husband, Sergey Kiselev, and I went to our favourite café to discuss the offer and stayed there until it closed, "Solovyova recalled." We decided: it was worth a try.

Irina was initially chosen to be the deputy to Valentina Tereskhova, the first woman to go into space on Vostok 6 in June 1963.

Solovyova was also chosen for the flight on Voskhod 5, in which the first woman would walk in space (as mentioned, that honour subsequently went to Svetlana Savitskaya over twenty years later, in 1984), but the Voskhod programme was cancelled after Voskhod 2, in favour of the Soyuz programme.

3.7   Tatiana Kuznetsova, the Emotional One

Kuznetsova was the only person in her family with a higher education. A member of the staff of the Moscow Institute of Radio Techniques, she quickly rose from the position of stenographer to that of party secretary at the Institute. A year later, she was promoted to senior laboratory assistant, although she did not yet have her degree.

Tatiana was one of the finalists, along with Valentina Tereskhova. She was in her early twenties and the youngest of the selected candidates. As part of the final test, she, Tereskhova, Irina Solovyova, Valentina Ponomareva, and Zhanna Yorkina (all aged between 22 and 28) had to enlist in the Soviet Air Force, specifically, within the Zhukovsky Academy of High Engineering. Despite being a favourite at the start of training, it was said that Tatiana was still "too young and sensitive" and had not fully developed "her qualities." She was also affected by the coincidence of an illness that kept her away from training for a while. She was not the only one to suffer that fate: Kamanin also withdrew Zhanna Yorkina for the same reason. But Voskhod-5 could have been Tatiana's second chance. This continuation of the Vostok programme included both manned and other direct flights from Earth. The Russian government's objective was clear: to once again confront the United States of America, which, by this time, was already beginning to develop Project Gemini.

3.8   Tatiana Morozycheva, the Artist

Tatiana Morozycheva was a woman who had style. She had worked as an art teacher in Yaroslavl while also enjoying skydiving. Morozycheva began representing her region in national competitions and helped Valentina Tereskhova in her efforts at the local skydiving club to which they both belonged. Her candidature was pre-approved by the local branch of the Communist Party.

Morozycheva competed with Valentina Tereskhova for a place on the women's space unit, but lost.

What happened next is still unclear.

One version of events says that Morozycheva married and became pregnant before being informed of her selection for screening, so she was not considered.

Another version says that she was outright rejected, and only later did it become known that she was expecting a child.

According to her friend Natalia Ledneva, who spoke to a local newspaper in Yaroslavl, Morozycheva was not an accommodating person. She spoke bluntly and strove to be number one. Ledneva recalled that Morozycheva, in the flat race, had been faster than her male counterparts, to prove that she was the best.

But the *Kommersant* newspaper suggested that Tereskhova had surpassed Morozycheva in something just as important to the Soviets as health tests: promoting communist values.

The all-female Voskhod flight was cancelled, first under pressure from Gagarin and the other male cosmonauts, and then eventually to focus on the development of the Soyuz space probe, which was deemed far more important. Moreover, according to the male chauvinist ideas of the time, every woman who flew in space took the place of a man.

3.9   Marina Vasiliyeva Popovich, or Madame MiG, Aviation Ace

Marina Popovich (born 1931, died 2017) was a Soviet Air Force colonel, engineer and Soviet test pilot. In 1964, she became the third woman and the first Soviet to break the sound barrier. Known as "Madame MiG" for her work, during her career, she set over a hundred world aviation records on over forty types of aircraft.

She was born in the Velizhsky district of Smolensk, but was evacuated with her family to Novosibirsk during World War II. She learned to fly as a child.

However, after the war, the Soviet Union banned women from serving as military pilots.

At the age of sixteen, she wrote to Soviet Marshal Kliment Voroshilov, claiming to be twenty-two and asking to be admitted to a flying school. Voroshilov stepped in to help, and she was subsequently admitted to the Novosibirsk Aviation Technicum, from which she graduated in 1951.

Initially, she worked as an engineer, and then later as a flight instructor. She and her husband, Pavel Popovich, both applied to the space team, and, as a result, she ended up joining the first group of women trained to become cosmonauts.

After two months of training, she was dismissed from the team. Pavel, however, was admitted to the men's program. In August 1962, he would be a part of the first group space flight, with Andryan Nikolaev, becoming the eighth person in space aboard Vostok 4.

After the last round of selection, Marina Popovich was told by the doctors: "*You failed the health tests.*"

However, it is unclear whether Marina Popovich's purported failure was actually true. Some documents relating to the selection process are still classified, and perhaps external factors were taken into account, including loyalty to the regime and some discriminatory assumptions. Later, all of the finalists would admit to feeling ill after each round of simulator training, but some were much better at hiding it than the "sincere" girls.

Some time later, Marina's husband would ask Kamanin to help her, but only to get her into the air force.

Marina became a pilot in the Soviet Air Force in 1963, and was admitted as a military test pilot in 1964.

In the same year, she broke the sound barrier in an MiG 21.

She entered the military reserves in 1978, and then joined the Antonov Design Bureau as a test pilot. There, she set ten flight records in the Antonov An-22 turboprop. She retired in 1984.

A star in the constellation Cancer is named after her.

She has published nine books and is a member of the Union of Russian Writers.

She is the author of a collection of poems, *Zhizn—vechny vzlyot* (*Life's An Eternal Rise*, 1972).

She is also the co-author of two film scripts:

- *Nebo So Mnoy* (*The Sky Is With Me*, 1974), and
- *Buket Fialok* (*Bouquet of Violets*, 1983).

She and Pavel had two daughters, Natalya and Oksana, and from them, two granddaughters, Tatiana and Alexandra, and a grandson, Michael, who was born in England.

After she and Pavel separated, her second husband became Boris Alexandrovich Zhikhorev, a retired major-general of the Russian Air Force and deputy chairman of the Central Committee of the Union of Soviet Officers.

The awards and honours granted to Marina included:

– the Order of the Red Standard;
– the Order of the Red Star;
– the Order of the Badge of Honour;
– Honoured Master of Sport;
– the Great Gold Medal "FAI" for the distribution of aeronautical knowledge.

Marina Popovich was buried with military honours in the Federal Military Memorial Cemetery, located in the Mytishchinsky District, Moscow Oblast, on the northeastern outskirts of Moscow.

Here is a professional portrait of the woman, as delivered to us by Wired magazine in 2017 (Italian edition), edited by Emilio Cozzi.

MARINA POPOVICH, THE LAST INTERVIEW.

She was among the best pilots of all time, a heroine of the Soviet Union. She passed away at almost eighty-seven years old. Wired Italia met her a few months before her death at her home in Star City. In her last interview, Marina Popovich talked about her extraordinary life. It is said that the sky is the limit. But the life of Marina Popovich,"Madame Mig," as she was called in the years of the Iron Curtain, has little to do with clichés.

After the end of the war, when there were only three women's aviation regiments, the Soviet instruction for female comrades was to give birth to future aces of the air, not to be part of them. Giving birth, but not flying. And to do "manly" work only in exceptional cases such as wars.

But Marina Popovich, a legend of aviation, a Socialist Hero of Labour and recipient of the Order of Courage awarded to her in 2007 by Vladimir Putin, thought differently. But when asked to introduce herself in the interview, she says something else entirely.

Ms Popovich, how would you define yourself?

"I am a poet, an engineer and a former first class pilot."

Let's remember that in 1950, she was not yet 20 years old. One morning she turned up at the Tušino aerodrome in Moscow with a declaration signed by Kliment Vorošilov, president of the Presidium of the Supreme Council of the Soviet Union. The letter orders the girl to be subjected to flight tests for admission to the air force. This letter had been obtained a few days earlier on Nikolai Kamanin's instructions.

How did it all begin? How did it go that morning in Tušino?

"After twenty-four months of refusals from the academy, I went in person to the airfield with Vorošilov's letter. And when I arrived, my blood ran cold."

Why?

"There were three Yaks on the runway for the exam, very different from the ones I had learned to fly in Novosibirsk. Fifteen minutes before the test, with the help of another candidate, I wrote down on a sheet of paper the piloting techniques of that unknown aircraft, the take-off speeds, the parameters of the aerobatic figures. Then, when General Balashov, the examiner, arrived, I asked him for a pillow."

A pillow?

"Yes. At 160 centimetres tall, I couldn't reach the plane's pedals. In response, Balashov asked me if I wanted a doll to make me feel comfortable. But I never had any and I told him so. (sic!) So I put the cushion down and took off. I did all the manoeuvres required, even pushed myself into a spin and then landed. A revisable re-entry, to be honest."

Did it go well?

"Balashov gave me top marks. "Especially since you've never seen this plane before," he said, startling me. I asked him how he knew: "If not, she would have fixed the pedals, it's adjustable," he laughed. From that moment on, I was a cadet in the Soviet Air Force. And the happiest girl in the world."

Why such determination?

"I wanted revenge and I made that clear to Marshal Vorošilov when I asked him for a letter of introduction."

Revenge?

"Everything I experienced came from the sky. Like the Nazi planes that in 1941 attacked Velizh, the place where we lived. I was only ten, but I'm sure they swooped down just to be more precise with the bombs. They razed whole villages to the ground. Then they occupied the whole area, rounding up and killing anyone who tried to escape. I remember when a plane killed a woman carrying water from the well. He opened fire twice, the second time when she was already on the ground. He wanted to make sure that she was really dead. At that moment I decided that it was from the sky that I would take my revenge. I realised that only from above, dominating the sky, could I protect my family. This is what I told Vorošilov; it was an obsession for me."

How did you come to enlist?

"After the Germans burnt down a village next to ours, my family decided to escape. We left for Siberia, staying first in the Ojash cemetery, as there was no other place for displaced persons, and two months later we arrived in Pushkarivka. From there we went to Novosibirsk."

Where you decided to enrol at the aeronautical institute....

"I actually enrolled in a welding school because the previous year, having caught malaria, my grades weren't good enough for aviation. When my name didn't appear on the list of those admitted to the flying courses, I freaked out. Only Igor Karpinsky, who was a well-known airman in the area, managed to calm me down, promising me that he would teach me to fly at the nearby aeroclub. He did just that. I remember my first solo take-off as if it were yesterday."

Your revenge would come true

"On the contrary, I distinctly remember the joy of flying alone, manoeuvring an aircraft as if it were a part of me. At that moment, the anger that I had been carrying for years, even after the war, suddenly vanished. It was replaced by a deep, solemn sense of responsibility for the people I had to protect. Always giving the best of myself. Raising my expectations day after day. That's why, once I became a pilot, my goal immediately changed: to test the best aircraft our engineers had built. I was going to be a test pilot. I made it eight years later, going from third to first class in 24 months. The highest."

What memories do you have of your long career as a test pilot?

"The fondest memories are not of me, but of my direct superior, Colonel Vasilij Gavrilovic Ivanov. He was a celebrity, you only had to say "VG" and everyone knew who you meant. Two stories would be enough to describe him. One day he was diagnosed with kidney stones, something that could get you banned from flights. He asked a motorcyclist to drive him over bumps and humps all day, running out of petrol a couple of times. In the evening the stone was gone. Another time he experienced engine failure a thousand metres off the ground. He could have ejected, but then no one would have known why he had failed. He decided to land on the undercarriage to save the black box. He cut through two electricity pylons and miraculously avoided the so-called "whiplash," the bounce that nine times out of ten destroys the plane. He then climbed out of the cockpit, took off his helmet and stuck a bunch of tulips he had just picked under the nacelle."

A gamble...

"No, he was a man who had regained his life. He brought those tulips to me: "*They're for you, Marina,*" he said. "*You can land a plane even with a non-operational engine. These flowers and I are the proof*." Once again it was the sky that changed my outlook."

Marina had to go further, to perceive the limit as a new starting point. In the 1950s and 1960s, experimental aircraft test pilots were dying at the rate of one a week. Marina Popovich's aircraft caught fire in flight twice. The first time, she was somehow able to survive the crash . The second time, she managed to land with an engine out of action and almost no fuel. A whole delegation was there to receive her: she had just beaten the world record, which was held by the Americans at the time, by two hundred and forty-four kilometres.

She had already met her future husband during her studies. He was also a cadet. On their first date, she had given him a bouquet of daisies, the aviators' flower. She picked them beside the runway in Novosibirsk. Pavel Popovich was so impressed by the gesture that, more than ten years later, on August 12, 1962, on Vostok 4, he would carry into orbit that very bouquet, which he had dried. Shortly after their marriage, Marina and Pavel were transferred to the fledgling Star City. Having started the recruitment programme in May 1959, they both passed the selection of the first astronauts. At that point, for Marina, the border she sought to reach became the stars.

Your job was one of the most dangerous. Were you ever afraid?

"A test pilot cannot be. His objective is to take the aircraft to the limit, to learn its secrets and explain them to others. You don't do it for yourself, but for those who will fly it later and for the people above whose heads you are flying. The priority is not your safety, but theirs."

And did you ever fear for your husband, once he was destined for space?

"We were sure that the engineers, especially after some accidents at the beginning of the programme, had everything under control. And they did. What would I have to fear anyway? A space launch was less complex than testing an experimental aircraft. Incidentally, I never failed to remind Pavel of one thing: he sang better than he flew. He had a beautiful voice."

Did you envy your husband the fact that he would get to be in orbit?

"On the contrary, I was proud of it. And I was sure I would follow him; both Kamanin and the chief designer of the space programme, Sergei Korolev, had not yet called on me because I was the wife of a cosmonaut, or whatever, I thought. I was a first-class test pilot, flying the best machines in the country every day. And to be honest, I don't even know if I would have given it up to do a few laps around the Earth."

What did you think when it became known that the first cosmonaut, in June '63, would be Valentina Tereshkova, an amateur parachutist and textile worker, and certainly not the best Soviet pilot? It turns out that you had passed the first tests with flying colours.

"I passed the first selection, but I was not admitted to the finalists announced in February 1962. No one ever explained to me why. Nikita Khrushchev himself is said to have decided: he thought it would be perfect to launch a woman worker with no particular flying skills into space. It would have demonstrated the technological and social superiority of the USSR."

Shortly after landing, Valentina Tereskhova was named a Hero, like Yuri Gagarin two years earlier; she was given a postage stamp and, in 1966, was admitted to the Supreme Soviet. How did you take it?

"I continued working peacefully. I also resumed my studies to become a university lecturer. Although the news was confidential, we knew that Tereskhova's flight had not been brilliant. Through no fault of her own, of course. But during the three days in orbit, she felt ill and was injured during the landing. The worst news was this: Korolev decided that no more women would be launched. That spread discontent."

The second cosmonaut, Svetlana Savickaja, took off only nineteen years later. Returning to you, is it true that it was the fact that you had a daughter that ruled you out? It seems that your husband and Gagarin did not want a mother to risk her life.

"If Pavel had anything to do with it, he didn't tell me. Kamanin never admitted it either. I think it was Yuri who objected... (*she pauses. Her eyes suddenly turn liquid, ed.*) Pavel? Who knows? Maybe... No, I think it was Gagarin."

A few years later, you and your husband separated...

"We still had some wonderful years. We just grew apart. Let's just say my job required a little less ease than his."

You're an aviation legend; do you have any regrets?

"No legend; there have been and will be better pilots than me. But I am proud to have become a first-class test pilot. There is a Japanese saying that I love: 'If you haven't encountered difficulties in your life, go out and buy them. Only by facing them can you become a complete person'. I survived *malaria, famine, war*; just before starting the academy, I almost couldn't fly because of frostbite. I was in the hospital for five months. Yet I have been flying all my life. Do you know what the secret is?"

Please tell me.

"Patience. That with commitment, you can go anywhere."

Is there really nothing you'd like to do that you haven't done?

"Go to the moon, look at my sky from an even higher place. And I don't exclude that I will do it. I'm patient."

3.9.1 Svetlana Savitskaya, the Only One Who Flew. Almost Twenty Years Later

The daughter of Yevgeny Savitsky, who was a World War II fighter ace, Svetlana showed a great passion for aviation from an early age. By her 22nd birthday, she had also recorded over four hundred parachute jumps and won first place in the World Aerobatic Championships. She obtained an engineering degree from the Moscow Aviation Institute in 1972, and was accepted as a test pilot. She eventually qualified to fly more than twenty different types of aircraft, earning a number of female speed and altitude records in the process.

In 1980, Savitskaya was selected for the Soviet space programme and began training. On August 19, 1982, as part of the Soyuz T-7 mission, she became the second woman to enter space. During her second trip (on Salyut 7), Svetlana became, on July 25, 1984, the first woman to perform a spacewalk, when she participated in welding experiments on the outer hull of the space station.

Savitskaya returned to Earth and took up an executive position in the Aerospace Design Office. She switched to politics and was elected to the Duma as a member of the Communist Party. She remained active in the Duma during the reforms of the 1990s. By 2003, she had risen to the fourth highest position in the ranks of the Communist Party, as deputy chairman of the Duma's defence committee; she was re-elected in 2007.

An all-female Soyuz flight (planned for Soyuz T-15 on International Women's Day in 1985) was later cancelled due to problems with the Salyut 7 space station. Only after the fall of the Soviet Union would Russian women begin flying to the Mir and ISS space stations as regular crew members, and not as mere objects of propaganda.

3.9.2 Male Colleagues

As we have seen, many qualified women participated—or at least attempted to participate—in the Soviet space program. This seems somewhat to speak to the stated lack of gender bias in Soviet society, but does it reflect reality? In the story of Valentina Tereskhova, which we will address in the next chapter, Tereskhova's male colleagues' thoughts on the matter were relayed by the Soviet cosmonauts themselves to their American comrades during the joint Apollo-Soyuz flight in 1975.

They felt it was a disgrace.

Both because a novice had been flown and because the reputation of Soviet cosmonauts (tough, brave men) had been tarnished. It had been shown that a young, unskilled woman could do the same things! In their partial defence, it should be noted that, some time before, many of them had trained hard for years to get the coveted place on board Vostok 6, and had been outbid, for purely political reasons, by the newcomer.

But there is further confirmation of the machismo of Soviet cosmonauts. In 1965, the flight managers of the Voskhod project planned an all-women mission. However, the male cosmonauts objected strenuously. According to them, the Voskhod capsules were more dangerous than the Vostok capsules (a fact that has been historically established), and consequently, in the event of unforeseen problems, at least <u>one man</u> would be needed to resolve the <u>situation</u>.

To support the cosmonauts, the space suit industry <u>refused to make a model specifically for women</u>!

In the end, the Voskhod project was abandoned in order to concentrate efforts on its successor, Soyuz. The pink expedition was not discussed again for some twenty years, when Moscow planned an all-female mission to the Salyut 7 orbital station.

This time too, however, nothing came of it: in 1985, the station suffered a major malfunction, forcing its occupants to make a sudden escape.

So, it was decided that <u>the most suitable people</u> should be sent to carry out the repairs, and the crew was replaced by <u>two male veterans</u>.

Shortly afterwards, Mikhail Gorbachev was elected to the leadership of the Communist Party; he felt differently about women being used as advertising frames.

However, in view of the deep economic and social crisis in his country, he considered space programs to be a waste of money, so the mission was permanently cancelled.

If we want to take stock of history, to date, <u>dozens</u> of American women have flown in space, while only <u>three</u> female cosmonauts have been on space missions, including Tereskhova. The other two are Svetlana Savitskaya and Yelena Kondakova. The former is the daughter of a high-ranking soldier in the Soviet army, with many supporters within the party. The latter is the wife of the astronaut Valery Ryumin, later promoted to senior executive of the Russian space agency, and also subsequently elected to the Duma. In short, on closer inspection, two "recommended persons."

On the other hand, there is another important detail: some of the women in the Soviet and then Russian astronaut corps left their posts prematurely (such as Nadezhda Kuzhelnaya), because they were tired of being outranked by their male colleagues who joined the programme after them. In 2020, the Russian cosmonaut corps numbered about forty men and only one woman, Yelena Serova.

As for the first "spacewalk" by a woman, the story goes that it was performed by Svetlana Savitskaya, but few know the grotesque background.

After her flight aboard the Soyuz T-7 in 1982, Savitskaya had retired from active missions.

But in 1983, NASA announced (aiming point-blank for the Soviets) that Kathryn Sullivan would perform an extra-vehicular activity in space the following year, lasting three hours and thirty minutes. The Soviets, not content to have the record taken away from them, hastily recalled Savitskaya to duty, put her through a quick workout, and launched her on Soyuz T-12 in July 1984, giving her a "spacewalk" of the highly suspect duration of …. three hours and thirty-five minutes. Three months later, Kathryn Sullivan carried out her planned three-and-a-half hour mission, and no one at NASA thought of extending operations by ten minutes to snatch a meaningless record.

To fully understand the mindset of the Soviet cosmonauts, a few things must be remembered. First of all, you need to know the anecdote of the fellow US astronaut who, while waiting for the launch, was asked by the control centre if he was nervous and replied:

> How would you like to be on board the joint production of a thousand companies that made the lowest bid at the tender auction?

In fact, one must remember the US disasters of the Challenger and Columbia shuttles, on which seven astronauts of both sexes died. But in the Russian arms industry, there were no low-bidding companies. There was only, or so it was said, the best.

Why, then, did the Russians have qualms about women? Here's the thing: on the occasion of the various disasters, in addition to due condolences for the loss of life, the Russians also expressed horror at the fact that the Americans had risked and sacrificed the lives of women. In their view, women should not have been allowed to risk their lives in such a dangerous job!

This idea was not new: Yuri Gagarin himself said that Valentina Ponomareva had been rightly rejected for the Vostok 6 flight because she was already a mother, and no mother should be allowed to risk her life in a rocket.

The most egregious discrimination against a mother, however, was used against Marina Popovich, the ex-wife of the Vostok 4 and Soyuz 14 astronaut. She had been in charge of testing dozens of the latest generation of military aircraft and held many records and awards. Marina Popovich applied to join the space programme, brilliantly passed all the theoretical and practical examinations, but was eventually rejected because she had a young daughter. According to some sources, even her husband opposed her application and tried to convince the selectors to reject her, which was one of the reasons for their divorce in 1968. But we read earlier that she never spoke about this in public. A true gentlewoman.

In any case, for future trips to Mars, doctors from the Russian space agency have already declared that their country's crews will be made up only of men.

To complete this picture of the role of women in space enterprises, it is worth mentioning the European situation. The ESA has always followed NASA's footsteps in the selection of its astronauts: the preferred candidates have always been pilots with excellent track records and psychophysical tests have been used as the main evaluation parameter. However, in the last selection of the astronaut corps, which took place a few years ago, a small innovation was introduced: the adoption of "pink quotas." In practice, since publication of the call for applications, it has been indicated that, in the event of equal qualifications, at least one place would be reserved for a woman.

4

VALENTINA TERESKHOVA, the Factory Worker Who Went to Space

Abstract The reasons for choosing Tereskhova for the first female flight into space are identified. We look at her hometown, her moral values, which were so in line with communism, her rebellious personality, her problems during the orbital flight (which lasted three days), and the hostility of Korolev and the medical staff. Moreover, some details about this flight are revealed, discovered after many years: the secret of why the spacecraft went in the wrong direction and that of the return to earth. It ends with the image that the West had of her and her daughter Elena.

4.1   Introduction

Having recruited the first two cosmonauts, Yuri Gagarin and German Titov, and launched Vostok 1 with the first man in space, Kamanin took a break.

But thereafter, he was hardly ever present at his home just outside Moscow, where he lived with his wife, son and granddaughter.

Kamanin neglected them largely to manage the Star City space team and pursue his dream of the first female spaceflight.

When the first cosmonauts travelled around the world to give speeches after their flights, Kamanin was with them. During these trips, he realised that one of the most frequent questions asked by foreign journalists was about sending a woman into space. This inspired Kamanin to go ahead with the idea,

© The Author(s), under exclusive license to Springer Nature Switzerland AG 2022
M. R. Menzio, *The Secrets of Soviet Cosmonauts*,
https://doi.org/10.1007/978-3-031-09652-5_4

says Anton Pervushin, author of *Yuri Gagarin: A Flight and a Whole Life* and *108 min that Changed the World.*

Kamanin was able to forge powerful alliances, including with senior party officials and, above all, Mstislav Keldysh, a member of the USSR Academy of Sciences and a leading scientist in the fields of mathematics and mechanics. Kamanin also sought the support of Sergey Korolev, who would prove to be a critical voice in the realisation of his dream.

After some effort, Kamanin managed to convince Korolev to support the idea of a first female flight. And six months later, the Central Committee of the Communist Party agreed to recruit sixty more cosmonauts, including five women.

What was required of the candidates? Apart from being experienced pilots or parachutists, five things were required:

- To not to be over 30 years old ("Sorry, madam, we want them younger.");
- To weigh no more than 160 pounds ("Sorry, madam, the scales say no.");
- To be no taller than one hundred and seventy centimeters. ("Madam, you should play basketball.");
- To have perfect health. ("Madam, that parameter excludes you.");
- To be of pure communist faith ("Madam, you come from a family that was close to the Tsar!").

Between March and April 1962, eight women were selected, after which five were chosen as finalists who went on to become regular army officers with the rank of lieutenant.

These women entered an industry heavily dominated by men. It was the early days of space exploration, which was still an unknown for humankind as a whole. When one of these pioneers, Valentina Tereskhova, returned to Earth as the first woman in space, the whole world celebrated a milestone for both cosmonautics and feminism.

In early 1962, the members of the men's space team gathered in a dining room in Star City and were joined by Yuri Gagarin. He announced to them:

Congratulations! Get ready to welcome the girls in a few days!

Cosmonaut Georgi Shonin recalls:

We, a small group of military test pilots selected for the space programme, lived together as one big family in Star City for two years. We shared the struggles and knew everything about each other, and now we had to accept new members of our family. [...] When we started training together, it was very unusual to hear female call signs like Cajka (seagull) or Bereza (birch), instead of Sokol (hawk) or Rubin (ruby). [...] But their intonations spoke for themselves. If a voice was sound, everything went as planned. But sometimes, their voices sounded pitiful. This meant that the instructor was working badly with them, and Bereza or Cajka was trying to solve the problem.

Decades later, Ponomareva recalled that:

The men treated us well, they helped us a lot and taught us how to solve theoretical and practical problems, as well as how to hide health problems. [...] But they were not very happy when we, five women, presented ourselves for the first time in Star City.

Korolev planned to fly two women aboard a multi-crew version of the Vostok, the Voskhod. This was to be a much more complex mission, involving both space piloting skills and the ability for one of the two cosmonauts to do a spacewalk. For this mission, Korolev would need a commander with Ponomareva's skills and an athlete with Solovyova's courage and strength. Thus, it was decided that two women would simultaneously pilot twin spacecraft into orbit.

Nikolai Kamanin believed that female cosmonauts should not lag behind their male counterparts. After cosmonauts Nikolai Andrianov and Pavel Popovich simultaneously flew two Vostoks in August 1962, a female group flight seemed the logical next step.

The dual female flight plan was approved by the entire Soviet hierarchy, but was scrapped at the last moment at a meeting of the Communist Party Presidium on March 21, 1963. Blame the party ideologist Kozlov and the head of the Ministry of Defence Ustinov, who said:

Only one woman will be able to fly, and for propaganda purposes.[1]

---

[1] An anecdote about French women comes to mind. Not everyone knows that, until very recently, women were not allowed to walk around the streets of Paris wearing a pair of pants. Unless the damsels were riding horses or bicycles. Unbelievable, but true. The story goes something like this.

An ordinance issued by the Paris Prefecture prevented ladies from dressing "like a man" until 2010. Too bad it was issued in 1799. At that time, the month of November was still called Brumaio, according to the calendar tradition of the French Revolution. And in the capital, it was decided that women could put on a pair of pants, if they really wanted to, but they would need a medical certificate, to be presented to the police, to get special permission.

The rule lasted until close to the present day. No one, or hardly anyone, noticed. In 2010, some Green and Communist city councilors presented two agendas to have the Prefecture's provision, which had lasted for more than two centuries, eliminated. In July of that year, Alain Umbert, of the Union for a Popular Movement, again brought the matter to the attention of the government.

The Ministry of Equal Opportunity confirmed the law's incompatibility "with the principles of equality between the sexes that are inscribed in the Constitution." The curious provision has also

At first, Irina Solovyova, Valentina Tereskhova and Tatiana Kuznetsova formed the leading trio. But as time went on, Kuznetsova was replaced by Valentina Ponomareva on the shortlist. Kamanin described Kuznetsova as the most easily influenced candidate, traits he did not consider ideal for a future national hero. But his main concern was Tatiana's health. She did not respond well to some tests. These included repeated sessions on simulators, which heated the human body to extreme temperatures and simulated the powerful gravitational forces of flight. Due to growing health concerns, Kuznetsova did not take her final exams in the autumn of 1962. The remaining four women received good marks and left the programme as licensed cosmonauts.

But Tatiana Kuznetsova was not the only person whose health was affected by the programme. Zhanna Yorkina had hurt her leg during a skydiving session and was forced to take a three-month leave of absence to recover. She was not able to catch up, and thus leave the programme with honour, insufficient achievement to become the first woman in space.

Meanwhile, a male cosmonaut (Bykovsky) was rushed to final training for Vostok 5, delaying flights for two months.

That left Irina Solovyova, Valentina Tereskhova and Valentina Ponomareva. Was it by exclusion that Tereskhova, the least able of the three, was chosen for the Vostok mission? Or was it perhaps Premier Krushchev himself who made the final selection of the crew?

Tereskhova embodied the qualities expected of the New Soviet Woman. She was a reliable communist, a worker of humble origins and a "good girl." But it was likely mostly this: she was of a proletarian origin. Ponomareva's background, in contrast, had opened many doors for her. She was a member of one of the most elite airclubs in the country (in Tushino). She was admitted initially to MIFI (one of the most prestigious colleges for math/physics education) and eventually graduated from MAI: the most well-known institute for preparing aviation engineers. Apart from her physical prowess, it was her education that made her the most qualified candidate for the space flight. Tereskhova had the lowest level of education out of all of the candidates. Tereskhova eventually was admitted and graduated from the Zhukovsky academy, but that happened only after the flight.

Let us also note one fact: at the time, Soviet cosmonauts were treated as national icons, and the trainees of the space programme were the next generation. The members of the space team were young, attractive, intelligent and

produced some effect in recent times. Until 2005, because of the provision, Air France stewardesses could not wear pants. Going back in time, it is enough to remember the opening in Parliament of a young congresswoman, Michele Alliot-Marie, then Minister of Justice. Blocked by the assembly, she managed to reach the Chamber only after much resistance and after having threatened to enter in her underwear, in order to eliminate the problem of pants.

well paid. The monthly salary of a cosmonaut was 350 roubles, almost three times that of a graduate engineer.

In a Soviet documentary, Kamanin admitted that he had been informed about Valentina Tereskhova only a few weeks before the official meeting. His deputy, General Goreglyad, had told him about her:

> We have a new candidate, and she is a very good candidate. She is a hard worker and a leader of the Komsomol! Please do not rush, we are still far from making the final decision on the flight.

According to Goreglyad, Tereskhova was best suited for the mission.

The phrase "*One day you will fly!*" was nevertheless said to all five women accepted for Star City's first female space unit.

However, the mission plan and launch date changed several times. At one point, the confusion was such that Kamanin was no longer sure of anything, not even that enough spacecraft would be produced in time for the flight. Among other things, the scientists had become aware of a new factor: for each cosmonaut, the radiation dose would far exceed the allowed quantity. It was a problem.

But by April 1963, the plan was in place.

Finally, the decision was made to fly a man, Valery Bykovsky, in one of the two Vostok spacecraft.

The question of which female cosmonaut would fly the mission remained open, while the search for the right person continued.

At that point, Kamanin began to worry about his "girls," as he called them. He remembered all too well the reprimands given to Gagarin and Titov for excessive drinking and reckless driving. As far as we know, the members of the female space unit never engaged in such behaviour, but some had their vices. Occasionally, Valentina Ponomareva smoked cigarettes, which was strictly forbidden; not only that, but she was known for her (moderate) consumption of alcohol.

According to Kamanin, the woman was also arrogant (a way of saying that she was forthright) and vain (perhaps she chose to dress somewhere between elegant and decent).

So, the man thwarted her.

He wanted a woman who was ideologically pure and who had completed parachute training of at least five or six months. He wanted "girls of strong communist faith" with experience skydiving in the aerodynamic circles of the Soviet Union.

When it came down to two finalists, there were only Ponomareva and Tereskhova. Who would be the first Soviet woman in space? Ponomareva had the best test results, but did not give favored answers in interviews with the communist selection committee, which was too puritanical and, perhaps, blinkered. The fact that Ponomareva did not have extensive experience with parachuting played a certain role as well.

4.2   The City of Yaroslavl

Whoever owns the Volga owns Russia.

Twenty kilometres from Maslennikovo, the village where Valentina Tereskhova was born, is the city of Yaroslavl, a place where the Trans-Siberian Railway passes through. As a child, when Tereskhova went to the city, she loved to watch the trains. How she would have loved to travel all over Russia, that endless country that she adored and that enchanted her!

In the city, she saw churches, temples, cathedrals, the whole of the old part with its onion domes, the symbol of the East and of Russia. She heard the friendly sound of the waters of the Volga and fell in love with the jewels of Russian architecture. Yaroslavl is one of the gems of the famous Golden Ring, eleven cities located between Moscow and the Volga. Valentina strolled along the banks of the river and arrived at the Old Town (a UNESCO World Heritage Site since 2005). She admired the Merchants' Lodge and the oldest Stable Theatre in Russia. She saw the monument to Prince Yaroslav the Wise. Legend has it that the prince killed a bear there and wanted to build a citadel. The city's coat of arms is a black bear holding a golden axe. On the same square, there is also the main entrance to the Spaso-Preobrazhensky Monastery, where the most important work of early Russian literature, the 'Tale of Igor's Campaign', was found. A one-thousand-rouble banknote can be your guide to Yaroslavl, as the most beautiful sights of the city are depicted on it. If you go through the Sacred Gate nearby and look back, you can see the view depicted on the banknotes: the chapel, a birch tree, the bridge, the gate and the tower. Yaroslavl is home to many beautiful churches, but the most beautiful of them all is St John's. It is not in the centre, but rather on the other bank of the Volga, and it appears as it it's wrapped in a Persian carpet. The steeple looks like a tower. The whole church is a masterpiece, the frescoes and decorations are ceramic. And its history is extraordinary, since it was built only with the offerings of the faithful.

Valentina was born nearby. As a tribute to her, a planetarium was named after her, a futuristic building with a telescope, a museum of cosmonautical history and a "star room" on whose domed ceiling images of the cosmos and films on astronomical phenomena are projected.

4.3   Communist Faith

> Politics is the choice between the disastrous and the unpleasant.
>   (John Kenneth Galbraith)

To the question:

What do you want from life?

Ponomareva replied:

I want to take everything it has to offer.

Tereskhova, on the other hand, said:

"I want to irrevocably support the Komsomol and the Communist Party." That was the right answer.

Ponomareva also claimed that a woman could smoke and still be a decent person. Not only that, she had taken scandalous unescorted trips—terrible! —within the city of Fedosiya while she was there for parachute training.

According to her health and readiness tests, Ponomareva might be the first choice for female flight, but her behaviour and conversations lead one to conclude that her moral values are not stable enough,

writes Kamanin in his diaries.
  Instead, Boris Cherkov noted:

A business session of the State Commission was held the morning of 4 June, and that evening a film and audio recording were made of a "show" session. Major Bykovskiy and Second Lieutenant Tereskhova were confirmed as spacecraft commanders. Male comments, unsuitable for audio recording, were inevitable.

"Look how Tereskhova has blossomed. A year ago she was a plain Jane, and now she's a real movie star," remarked Isayev seated next to me. "And imagine what she'll be like once she's flown," I responded, and we both knocked on our wooden chairs. True, after getting a look at Ponomareva, we decided that she "didn't look half bad" either. But she didn't glow like Tereskhova. Judging by her appearance, she was overly serious, and it seemed to me that she was simply pouting because she remained the backup.

Ponomareva's memoirs paint a different picture. She recalls being excited about her role on the space team and working hard to succeed.

Unfortunately, she was the only woman without much skydiving experience, and on one jump, she landed incorrectly, injuring her tailbone. She could barely walk, but chose to jump again to overcome her fear. This second attempt was no better than the first, and her instructor was forced to call a doctor. All X-rays taken of cosmonauts had to be reported to the Kremlin, which meant she would be at risk of dismissal. Her doctor eventually decided not to take the X-rays, hoping that nothing serious had happened, and Ponomareva was grateful for his discretion.

This was not unusual: fearful of losing their prestigious positions, space crew members (both men and women) tended to hide medical problems, including minor illnesses. Decades later, Ponomareva discovered three cracks in her spine and one in her chest, all caused by failed parachute jumps.

Ponomareva recalls that there was no envy among the women of the team. According to her, there was a healthy spirit of competition. They all did their best to be number one, but also to support each other's efforts.

4.4   Buying Panties

Many women on the team described Valentina Tereskhova as a good friend. In fact, Zhanna Yorkina recalls:

> She always defended our interests in front of the bosses. For example, at the beginning of the programme we lived as if we were behind barbed wire. We lived near Moscow, but only Muscovites were allowed to leave the training camp to see their families. [...] Tereskhova and I were bored and asked permission to go to Moscow. What for? What do you want to buy? they asked. Once, Valentina Tereskhova lost control and said: We want to buy panties! - and in this way we got permission.

Every day that passed was further proof that Ponomareva had more skills and competences, but Kamanin felt that she was not suited to working in a group, that she showed too much independence and self-confidence. In short, he disliked her. With Tereskhova, Kamanin killed two birds with one stone: he ruled out Ponomareva, who was too sophisticated for him, and chose the iron communist, which pleased Krushchev.

Valentina Tereskhova was, in fact, attracting a lot of attention, and it was soon officially confirmed that she would fly, with Ponomareva and Solovyova as reserves.

Valentina Ponomareva, Irina Solovyeva, and Valentina Tereskhova (free From Boris Chertok's archives)

According to other sources, it was Yuri Gagarin, commander of the detachment, who influenced the choice of the first astronaut. Yuri had spoken out against Ponomareva from the beginning, a fact that she herself recalled:

Among the members of the commission was Yuri Gagarin. We went into the hall, answered a few questions and then waited for the verdict in the corridor. They told us that Zhanna Yorkina and I had passed. Some time later I heard the confidences of the deputy head of the Cosmonaut Training Centre Nikolai Nikeryasov. He confirmed to me that Gagarin had opposed my application. He said: for reasons of space exploration, you can risk the lives of male pilots if you really need to, even if it would not be worth it. But it is unacceptable to risk the life of a mother.

Instead, Gagarin threw his support behind Valentina Tereskhova, having noticed her among the others. He probably liked her strong-willed character. Or maybe he just liked the fact that she was single.

In any case, everyone knew that the chosen one would write her name in history, while the others would have to be content with the much more modest role of unknown participants in a historical event. The decision was therefore crucial: it had to be decided who would fly on Vostok.

Korolev had two separate conversations with the women who would act as Tereskhova's reserves after the decision had been made. For her part, Solovyova was told:

> We need a more outgoing person, because after the flight she will have to deal with publicity all over the world.

Basically, he told her: "You are worth nothing as a communicator."

Tereskhova had certainly proved her affinity for handling publicity. Most of her life, she held various public office jobs, including being a member of the Duma.

For Valentina Ponomareva, the excuse was different. Korolev told her:

> A working-class woman will be a better representative of Soviet ideals: you come from a white-collar family.

Classism in reverse.

> I have no doubt that Ponomareva was the most suitable for the first female flight,

says writer and space historian Anton Pervushin. But he also says that, unlike in Gagarin's case, the final decision was made not by specialists, but by high-level politicians. In fact, among these was Krushchev, who was looking for a 'Gagarin in skirts.'

Krushchev believed that Tereskhova would be a better representative of the ideal Soviet woman not only because she was a hard worker, but also because the textile industry that she represented played a key role in domestic politics. In short, a choice dictated by both reasons of state and reasons of industry.

All three women in the finals followed the same standard procedures before the launch. They filled in the captain's logbook, checked their spacesuits and got used to the spacecraft's cabin. But by this time, Ponomareva had lost all motivation, and there were moments when she couldn't stop herself from crying.

But the fact was that all of the stars had aligned in favor of Valentina Tereskhova. She was an outstanding parachutist. She had proletarian origins. Plus, her war hero father, who may have been killed in action, also played in her favour. All of this contributed her luck: her communist faith, her determination and her family.

Valentina's mother had told her that, during the Stalinist period, life was condensed into two "L"s: "Lack" and "Lines." That is to say, starving and queuing for food. She wanted to change her fate. She was one of those women who said to themselves: "If you want something to happen, then make it happen."

The women began their training, but this had to be a secret. Tereskhova did not tell her family what she was doing: she reported that she was taking part in special parachute training. The initial Vostok programme was changed so that a man, Valery Bykovsky, would be launched aboard Vostok 5, and, two days later, a woman would be launched on Vostok 6: this was Tereskhova herself, with her two reserves Solovyova and Ponomareva.

Valentina's path was already pointing towards history.

The joint mission began on June 14, 1963, with the launch of Bykovsky on Vostok 5 from the Baikonur cosmodrome in present-day Kazakhstan. After that, Tereskhova took off. But her flight was not without criticism.

Ponomareva took issue with the criticism of Tereskhova. She wrote:

I have no doubt that she did everything she was supposed to do, because we needed to understand how a human being would feel in orbit. The first six cosmonauts had no more important objective than this. All the scientific experiments in orbit were also important, but not crucial.

Very generous: she didn't think Valentina had taken her place!

The rest of the women's space unit continued to prepare for the next flight, trusting Korolev's word that they would all make it into space one day. Kamanin tried to convince Korolev of the value of a women's group flight, but there was no good reason for the Soviets to pursue this goal. There was no political necessity, as Tereskhova's flight had already provided enormous propaganda value.

Korolev died in 1966, and the next two years brought the deaths of two famous cosmonauts: Komarov and Gagarin. These accidents suspended the entire space programme, and the female cosmonautics unit was disbanded by 1969. Kamanin, unable to get his women's team off the ground, retired in 1971.

After their dismissal from the space team, each woman received a comfortable flat from the government, and the legacy of their cosmonautical training continued to have a lasting impact on their personal lives. Following the programme, every former team member married another cosmonaut. Four out of the five women remained in Star City and continued to work in the space industry. All files relating to their training programme would remain classified until the 1980s.

All of the women in training, with the exception of Tereskhova, were forbidden to become pregnant until the space team was disbanded. Ponomareva, who gave birth to her son before joining the programme, also had to obey this rule. Yorkina broke the agreement: as a punishment, she did not receive the military rank granted to all of the other women trained as cosmonauts.

Valentina Ponomareva went on to obtain her PhD, and played other roles in the Soviet space industry. After the collapse of the USSR, she returned to literature, writing several books about her time in the space team. In later years, she devoted herself to space psychology and developed recommendations for future cosmonauts.

Tatiana Morozycheva, considered for the space team but not accepted, gave birth to a son and continued her record-breaking career in skydiving. When she retired, she joined a local arts foundation and earned a living working for private clients. Morozycheva had problems with alcoholism that contributed to her death, despite interventions by Tereskhova, who remained close to her.

4.5   Thus Spoke CICAP

According to CICAP (the Italian Committee for the Control of Affirmations on Pseudosciences), "The alleged greater open-mindedness of the Soviets towards women is a myth, constructed through a skilful propaganda operation."

As Nicola Boschini states, Krushchev had undoubtedly deftly constructed his publicity campaign, as the flight of Vostok 6 became one of the milestones in the history of astronautics and influenced Westerners' view of the Soviet space programme. At that time, many US congressmen and their wives were outraged that women's emancipation in a communist regime was much more advanced than in their own country. Even today, in some ways, women's emancipation is more advanced in Russia than in the US.

There is a rumor that is denied in recently surfaced historical documentation, but that Kamanin's diaries confirm. According to these rumours, the marriage between Tereskhova and her colleague Nikolaev, the pilot of Vostok 3, was also one of Krushchev's propaganda ideas, based on the fact that both were very famous people at the time. Suspicions arose for the following reasons: all but one of the female astronaut candidates were unmarried, and the only unmarried male astronaut (Nikolaev) had been chosen to train them (together with Gagarin). What is more, the woman who, a few weeks earlier, had officially become engaged to her instructor was chosen for the flight.

Kamanin held Tereskhova in high esteem because of her strong yet sociable personality. Moreover, Gagarin and Titov had undermined their reputations as two of the most famous ambassadors of communism in the world by turning to alcohol, women and parties after becoming famous. So, Kamanin took great account of the personalities of those who would fly, believing that Ponomareva was an exceptional pilot but difficult personally. Kamanin was an old-fashioned Stalinist and a man of deep-seated chauvinist convictions: he immediately took a dislike to Ponomareva because she was self-confident, independent and very refined.

What an incredible hoax the years following Valentina's flight were!

Instead of advancing one step further, there was no trace of the feminism shown up to that point in Soviet cosmonautics.

Everything went back to the way it was, only worse than before.

Moscow shelved the women's cosmonautics programme for two decades.

4.6   Valentina's Life

A person who dares to waste even one hour of his time has not discovered the value of life (Charles Darwin).

Valentina Tereskhova was born on March 6, 1937.

At first, no one could have predicted how great a part Valentina would play in Russian history. She was of working-class origin, with a father, Vladimir Tereskov, who was a tractor driver, and a mother, Yelena, who was a textile worker.

There was also an older sister, Lyudmila, and a younger brother, Vladimir. The family lost their father, who fell in the Soviet-Finnish War, when Valentina was two years old.

After the world war, everyone moved to Yaroslavl, where Valentina completed her primary education. From an early age, she showed a good ear for music and learned to play the domra (a Russian musical instrument similar to the lute, considered the mother of the Balalaika). However, the girl had to help her mother, as there was never enough money. Valentina started school in 1945, at the age of eight, but left in 1953.

She continued her education first through correspondence courses, dividing her time between work and books. She went ahead with her high school education by attending evening classes after the workday, and, in 1955, completed her early studies. She had an unshakeable will, attending evening classes and graduating as a technical expert in 1960. As a worker, she first found employment in a tire factory, where she operated a diagonal cutting machine, then in the industrial textile factory in Krasny Perekop where her mother and older sister worked. For some time, she was also a seamstress and ironer.

However, in the Soviet workforce, Tereskhova was more than just an average worker.

First of all, she was very ambitious.

She learned to scuba dive through the DOSAAF Aviation Club in Yaroslavl, an auxiliary organisation of the Soviet Air Force. Kamanin was one of its founders. DOSAAF stands for "Voluntary Association for Cooperation with the Army, Air Force and Fleet." The stated aim of the society was 'the patriotic education of the population and its preparation for the defence of the Fatherland.'

Among the means to achieve this goal was the development of paramilitary sports such as parachuting. This is where Valentina's preparation comes in, as she made her first jump on May 21, 1959, at the age of 22.

We can imagine how she felt at that moment (from an interview with a skydiver):

I'm in the plane and I see the lake, the woods, the rivers. It doesn't seem possible, the city is so small, the streets, the houses, everything is like a miniature. I feel the wind on my face. I am halfway out of the plane. My feet are hanging out. I lean against the door, the instructor tells me that this is the right moment, pats me on the shoulder and I push myself out into the void. Under my feet, everything flows. I see the lake, it's far away. I try to balance my arms, to keep the position, the air tries to turn me upside down, my feet go in all directions, I see the ground approaching, I open my mouth in amazement, but I close it again immediately. I fall, I'm afraid of ending up upside down, of not finding the strap at the right moment, of crashing, but the adrenalin is starting to take effect. I start to feel strong, in tune with nature, with the air currents. I see the details. I am flying. One thousand eight hundred metres, I always look at the altimeter, then one thousand seven hundred, one thousand six hundred. Now! I grab the strap, pull it, and like a missile I am shot upwards, I see that the parachute opens, all the wires are stretched and the sky disappears back there. I start to descend again, but this time in a controlled, slow way. I am suspended in the sky, I see the treetops, below me the golden domes of the churches of Yaroslavl, the sails of the ships. I feel like a seagull, suspended in the sky, looking for the right moment to glide elegantly towards the sea. I orient myself towards the landing point. I go down, to the ground, hold on to the straps, fall, slowly, like a snowflake. Here I am, now I recover everything. It was magnificent.

She became a good skydiver, performing over ninety jumps. Little did she know how much it would change her life.

She later founded the "Textile Mill Workers Parachute Club" and acted as its first leader. She also became certified as an expert in cotton spinning technology. Two years later, in her factory, she became secretary of the local Komsomol (Young Communist League), an organisation that was, in spirit, the youth division of the Communist Party. Being the leader of that sector was the opportunity that opened the door to space for Valentina.

Valentina was a great admirer of Gagarin. After the cosmonaut's debut, she tried several times to apply to the school for aspiring cosmonauts, like thousands of young people from all over the Soviet Union.

Her application was rejected.

But what she did not know was that Korolev and Kamanin were starting to recruit women to train as cosmonauts, and that the idea of a female cosmonaut also excited Nikita Krushchev, who recognised the propaganda value of the idea.

This would have demonstrated, firstly, the reliability of Soviet spacecraft and, secondly, the capabilities of female workers, whom the Soviets considered equivalent to males. At least, in theory.

On February 19, 1962, Valentina fulfilled her dream and took the entrance examination: she was promoted and accepted for training as a cosmonaut.

Tereskhova met the requirements and was chosen, despite the fact that her father was listed as MIA ('Missing in Action'), as opposed to KIA ('Killed in Action'). This raised the possibility that the father had escaped from duty.

However, her credentials as a Komsomol leader allowed her to overcome this obstacle. On the scales, her personal licence as a communist leader was worth more than the doubts about her father's death.

Tereskhova was therefore one of the five women selected (initially, one of the more than 60, but ultimately one of the five finalists). She was the least qualified of the candidates and was not highly educated. The other four women included test pilots, world-class parachutists and an engineer.

She was trained in body resistance to the factors of space flight.

She was trained in a heat chamber, where she stayed in a flight suit at a temperature of +70 °C and 30% humidity, then in a sound chamber (sound-proofed), where she was to spend ten days. Her zero-gravity training took place in a MiG-15.

Valentina Tereškova's family only found out about her mission after the fact, by listening to the radio. The flight could have ended in tragedy, so Valentina kept her family in the dark: it was not her decision; it was requested (read: commanded) that she do so.

At the time when Tereskhova was appointed pilot of Vostok-6, she was ten years younger than Gordon Cooper, the youngest American astronaut.

After the flight, Valentina continued her studies: she obtained a degree in mechanical engineering with top marks from the Zhukovskiy Air Force Academy, which she attended from 1964 to 1969. Tereskhova subsequently wrote several scientific papers.

From April 30, 1969, to April 28, 1997, Tereskhova was an instructor-cosmonaut-tester for the Orbital Complexes Group.

In 1995, she was awarded the rank of Major General (the first woman in Russia to be so honoured). On April 30, 1997, due to reaching the age limit, Tereskhova left the team. Apparently, she was the last of the original five women cosmonauts to have done so.

One of her greatest passions has always been motorsport, especially sports cars. In 1967, she asked to visit the Alfa Romeo factory in Arese, where she drove a GT Junior that had been made available to her for the occasion!

In the meantime, she was thrust into a hectic life as an important communist politician and international representative. She was called "Miss Universe," and poems and songs were dedicated to her.

After her retirement from active political life, Tereskhova continued to participate occasionally in space demonstrations and public events.

Today, she is a prestigious and honoured figure, among the major players in the history of human space exploration.

She is an honorary citizen of several cities in Russia, France, Great Britain, Italy, Bulgaria, Slovakia and Mongolia.

In Moscow, in 1963, she married cosmonaut Andrian Nikolaev, twice a Hero of the Soviet Union, who had participated in the Vostok-3 mission. The lavish celebration was led by Head of State Nikita Krushchev himself, who wanted and organised the union between the two cosmonauts. Apparently (but only apparently), the scientists were quite eager for the match-up as well, having decided to conduct an experiment: to see what children born of two people who had travelled in space would be like.

These speculations were always angrily rejected by the couple.

The wedding was nevertheless used for Soviet propaganda purposes. In 1964, their daughter Alenka, or Elena Andrianovna, was born, the first person to be born to two cosmonauts. But the marriage was not a happy one, and Krushchev's "space family" disintegrated within a few years.

Kamanin was constantly confronted with their disputes.

Valentina's husband, Nikolaev, was the third Soviet cosmonaut. Burly in personality, he always had time for his male friends, but never for his wife. In an interview, Tereskhova said that it was great to work with him, but that he had become a tyrant at home.

However, as was the case with American astronauts at the time, divorce would have meant the end of their careers. So, the couple stayed together, even after she met Yuliy Shaposhnikov, a handsome general in the medical service, director of the Central Institute of Orthopaedics and Traumatology. The two fell in love, but it was not until 1979 that Tereskhova separated from Nikolaev. However, her request for a divorce had to wait for the personal permission of the Soviet Premier Brezhnev, which only came in 1982.

Nikolaev never married again. People who knew him said that he did not want to share his life with any woman but Valentina.

Tereskhova and Shaposhnikov lived happily together for twenty years, until Juri Shaposhnikov passed away in 1999, leaving Valentina a widow. After the death of her life partner, Tereskhova retired to a small brick dacha on the outskirts of Star City. The house is topped by a seagull aviary, commemorating the call sign that she chose for her space flight. Valentina enjoys visits from the friends that she has collected over the course of a long and full life.

She loves her daughter and grandchildren Andrei and Aleksei very much.

In 1994, Valentina was appointed director of the "Russian Centre for International Cultural and Scientific Collaboration."

Her governmental appointments are uncountable.

She has always been a leading member of the Communist Party and a representative of the Soviet government in numerous international women's organisations and events.

She was a member of the World Peace Council in 1966; in the same year, she was called to join the "High Soviet of the Soviet Union."

In those years, she became an important female politician, working on international councils and speaking at cosmopolitan conferences. She held various political posts until the dissolution of the Soviet Union.

She took an interest in women's issues and took on many responsibilities,– the full list of which is too long to include here, so only the most important posts are listed below:

- Chairman of the Women's Committee of the Soviet Union,
- in 1971, member of the Central Committee of the USSR,
- since 1976, vice-chairman of the Commission for Education, Science and Culture,
- in 1967, member of the Supreme Soviet in Yaroslavl,
- from 1966–1970 and 1970–1974, member of the Council of the Soviet Union,

- elected to the Presidium of the Supreme Soviet in 1974,
- represented the Soviet Union at the UN Conference for the International Year of Women in Mexico City in 1975,
- continued as a deputy to the Supreme Soviet, Vice President of the International Women's Federation and many other international positions through the 1980s.

Among her official certificates and honours are:

- two "Orders of Lenin,"
- one "Golden Star Medal,"
- a "Joliot-Curie" Gold Medal,
- the highest honorary title in the USSR, the medal of "Hero of the Soviet Union,"
- the "Golden Peace Medal of the United Nations,"
- the "Simba International Women's Movement" award,
- a "World Connection Award," presented to her in Hamburg in 2004 by Nobel Peace Prize winner Mikhail Gorbachev.

In 1983, a commemorative coin featuring Valentina Tereskhova was issued, making her the only Soviet citizen to have her portrait on a Russian coin during her lifetime.

There are schools, sports centres, monuments, and songs, as well as streets in some 30 cities, dedicated to Valentina.

A lunar valley is named in her honour: "Tereskhova Valley."

Tereskhova's image has often been that of a primitive communist, perfectly integrated, but cold and with an iron will. Personally, however, she always felt that all of the obligations of the party were an unfair burden.

She always dreamed of going back into space, in particular, being part of the first expedition to Mars. The dream of going to Mars was shared among the first cosmonauts. Some were even willing to go on a one-way mission. In an interview, Tereskhova said.

Mars is my favourite planet, and it is my dream to get there to find out if life ever existed there. And if it did, and why it disappeared.

To fulfil this dream, she too was prepared to fly to the red planet on a one-way trip, without the possibility of returning to Earth.

Moreover, although she has been called the "iron lady" by some, Valentina's tireless later work amply demonstrated her humanity: from relentless assistance to troubled citizens to personal support of many orphanages.

In modern Russia, Valentina, a retired cosmonaut, was elected to the State Duma, the lower house of the Russian parliament. She was also vice-chairwoman of the Duma's Foreign Affairs Committee.

Valentina has continued to travel the world to speak about her experience in orbit, trying to raise awareness among the younger generation. To Americans, Asians and anyone who saw her, she said the same thing, which is how incredibly beautiful the Earth is and how important it is to take care of it.

Our planet is suffering from human activity, from fires and from war: we must save it.

On the 50th anniversary of her pioneering space mission, Tereskhova received one of Russia's highest honours, the Order of Aleksandr Nevsky, from President Vladimir Putin.

With seven other Russian personalities, Valentina participated in the 2014 Winter Olympics in Sochi, in the opening ceremony, waving the Olympic flag.

On March 16, 2020, Russia's Constitutional Court gave the green light to Putin's reform. This included the amendment proposed by Valentina Tereskhova herself, which reset Putin's number of presidential terms to zero, allowing him to run for the Kremlin again in 2024.

It was not until the late 1970s, with the impending flights of American women on the space shuttle, that the Soviet government recruited yet another group of cosmonauts. In 1978, Tereskhova and Kuznetsova applied for a new Soviet training programme. Both passed health tests, but were eliminated for being over the age limit.

Valentin Glushko, who had headed the space design office, had once promised Air Marshal Evgeny Savitsky that he would send a younger trainee, and somehow Savitsky's daughter Svetlana was chosen.

But Glushko wouldn't keep his word until two decades later: in 1982, the year Kamanin died, Svetlana Savitskaya would become the second Soviet woman in orbit.

The first American woman would not fly until June 1983. Twenty years after Valentina.

4.7 The Flight

When you walk on earth after flying, you will look up to the sky because that is where you have been and that is where you want to return. (Leonardo da Vinci).

Valentina during the flight: free photo from ESA

In the Baikonur cosmodrome, the Russians celebrated one success after another. A glorious ensemble of comrades, a great people wrote a new chapter in the history of mankind in June 1963.

June 14, 1963: three hours behind schedule, **Vostok 5** was launched into space. On board was cosmonaut Valeri **Bykovsky**, call sign Oak. Almost immediately, it became clear that the height of perigee (one hundred and eighty-one km) was lower than calculated: on the eighth day of flight, the spacecraft would only be able to make an uncontrolled descent. For this reason, Vostok 5 returned to Earth after only five days. In the meantime, Bykovsky quickly adapted to the weightlessness, and conducted observations of the Earth, the Sun and the stars. He also conducted scientific experiments: for example, he monitored the growth of pea plants under spaceflight conditions.

Two days later, on **June 16, 1963**, the **Vostok 6** spacecraft was launched into space. It orbited the Earth and began another joint flight; but this one was special. On board was **Valentina**.

Earth heard the voice of the person on Vostok 6.

The voice spoke. It said, "*Hello, I'm Cajka.*"

And immediately the "*Oohs*!" of wonderment began.

It was a woman.

It was Valentina Vladimirovna Tereskhova, who made history that day as the first female cosmonaut in orbit. As commander of a spacecraft, it was up to her to choose a code name for the radio links, and she chose the appellation Cajka ('seagull').

On that day, Tereskhova had confidently approached her Vostok 6 space-craft. As she reached the cabin, the historical significance of the moment had sent her adrenaline racing.

> She is well prepared for the flight. Not only will she be flying in space, she will be guiding the spacecraft in the same way as men. When she lands, we will compare who is better at completing tasks.

These are the words that Yuri Gagarin said in the Baikonur cosmodrome, a few hours before Tereskhova's launch.

In fact, Kamanin said that Tereskhova's start reminded him of Gagarin's. Like April 12, 1961, everything began perfectly on June 16, 1963. The conclusions of both those who watched Tereskhova as the ship was launched into orbit and those who heard her reports on the radio were unanimous: *"She had a better start than Popovich and Nikolaev."*

"Yes, I am very happy that I was not mistaken in my choice of the first female astronaut," Kamanin said at first.

Before the flight, she said: *"Hey! The sky! Take off your hat!"* (a quote from V. Mayakovsky's poem "A Cloud in Pants").

Valentina was the sixth cosmonaut in the USSR and the tenth in the world.

She established a long radio connection with Bykovsky, had a conversation with Krushchev and reported intelligently on the flight. The spacecraft's orbit had an apogee of two hundred and thirty-one kilometres, a perigee of one hundred and eighty kilometres and an orbital period of about eighty-eight minutes. The two ships (Vostok 5 and Vostok 6, respectively with Bykovsky and Tereskhova) had different orbits, but were to approach each other peri-odically, and for short periods, within a three-mile distance, allowing the two cosmonauts to communicate directly several times.

And all seemed to work as planned!

It was an outstanding achievement.

Tereskhova was not only the first woman in space, but in her first (and only) launch, she "collected" more flight time than all previous US astronauts combined (70 h and 50 min in space, with 48 Earth orbits). In space, among other tasks, Valentina carried out experiments designed to study the effects of microgravity and being in the cosmos on the human body.

The cosmonaut flew a total of some 1.97 million kilometres.

4.8 One Trouble Leads to Another

The flight was not without problems, however, and it was said that these were Valentina's own fault.

On the second day, June 17, flight leaders drew attention to Tereskhova's responses that day, which sounded evasive to them. Later, in her report to the State Commission, she said that, on the first day, she had hardly felt the spacesuit, but then her right leg had begun to hurt, a pain that did not cease until she landed. Valentina was plagued by nausea and could not eat the supplies that were prepared. She could not conduct any more scientific experiments, as she could not free herself from her sitting position, nor could she reach the materials.

The State Commission decided to land Bykovsky's spacecraft at the 82nd orbit (by the end of the fifth day) and Tereskhova's spacecraft at the 49th (by the end of the third day). On June 18, Tereskhova tried manual control, which could be used in the event of failure of the automatic orientation system prior to descent.

However, she did not succeed on either the first or the second try.

Everyone on the ground was worried.

They turned on the camera and saw that she was asleep. They woke her up, and told her about the upcoming landing and the manual orientation.

She replied: "*Don't worry, I'll do everything in the morning.*"

She would take a break during the night.

On the morning of June 19, during the 45th orbit, Gagarin personally read the instructions to Tereskhova, asking her to confirm each step. She turned the ship around, doing everything as planned.

But many accusations were still later levelled at her: that she had had trouble adjusting to gravity, and that she had experienced nausea and vomiting, and that she had been restless for most of the flight, and that she had complained of various pains and hygiene needs. The latter had not been resolved during training, because there was no time: Valentina had to enter the History of Space Enterprises at once. Right away, that is, before the Americans.

Sergei Korolev wrote in a diary on June 16, 1963:

I have spoken to Tereskhova several times. She feels tired, but does not want to admit it. In the last communication session, she did not answer her calls. We turned on the camera and saw that she was sleeping. I had to wake her up and talk to her about both the impending landing and the manual orientation. She tried twice to orient the ship and she honestly admitted that she was not successful in pitch orientation. This is a matter of great concern to us.

Some details of her problems can be found in Boris Chertok's memoir, *Rockets and People: Hot Days of the Cold War*, which we will discuss later:

We, the engineers who designed the control system, believed that it was much easier to control a spaceship than an aircraft. All the processes are more extended in time, and there is an opportunity to think. The spaceship will not fall on its tail, therefore, according to the laws of celestial mechanics, the spaceship cannot leave its orbit. Therefore, it can be controlled by any physically and mentally normal man with two or three months of training, and even by a woman!

According to Professor V. I. Yazdovsky, head of medical support for the Soviet space programme, women were less likely to tolerate extreme spaceflight loads from the fourteenth to the eighteenth day of their monthly cycle, i.e., during menstrual flow. Doctors therefore recommended that women not be sent into orbit on those particular days. However, due to the delay in the launch of the vehicle that carried Tereskhova into orbit, and the heavy psychological and emotional burden of the launch itself, the flight regime prescribed by the doctors could not be sustained.

So, did Valentina go into space during her menstrual period? She never denied it, nor did she admit it. Yazdovsky also notes that:

With Tereskhova, communication with earth has been slow. The woman is severely restricted in her movements. She sits almost motionless.

Despite nausea and physical discomfort, Tereskhova was able to endure forty-eight revolutions around the Earth and spend nearly three days in space. She kept a logbook and took photographs of the horizon. However, according to some detractors, the logbook was only finished once she returned to Earth.

Alone, therefore, Valentina was unable to manually manoeuvre Vostok's orientation to re-enter the atmosphere. She only needed the assistance of the mission control centre, from which Gagarin radioed the cosmonauts in space.

She returned to Earth on June 19.

After being automatically ejected from the capsule, she made her descent using a large heavy parachute that opened at an altitude of four kilometres.

She came ashore some six hundred and fifty kilometres from the intended target, which was the city of Karaganda in Kazakhstan. In fact, she descended near Novosibirsk, in the Baevsky district of Altai Krai.

Vostok 5 had re-entered the atmosphere shortly before. Bykovsky, automatically ejected from the spacecraft after re-entering the atmosphere, landed by parachute a short distance from and about three hours after Tereskhova.

Vostok 5 had been programmed to spend a record eight days in space, but Bykovsky had had several problems, in addition to the perigee being too high than expected. There was also a problem with the thermal control system. For these reasons, the team on Earth landed him after only five days, only three hours after Vostok 6.

A few days after her return, the Government awarded Valentina the high honour of "Cosmonaut-Pilot of the Soviet Union."

She returned to Earth as a world celebrity, receiving a large number of awards. The Soviet leadership had no doubt that this historic flight was a great political victory that would help promote communism worldwide.

> The Communist Party, the Fatherland and the great people of the Soviet Union have given us strength and wings to make this flight,

said Tereskhova, standing on Red Square between Krushchev and Yuri Gagarin.

> The fatherly words of Nikita Sergeevich [Krushchev], in a conversation we had on the first day in orbit, spurred me on to be brave.

Tereskhova then met with the assembled Muscovites, with Bykovsky, saying:

> Flying over all the continents, my celestial brother Bykovsky and I had a good time.

The words had been written by good publicists, and it shows.

Moreover, the celebration was carefully planned in advance: no detail could be overlooked, including the portraits of Valentina Tereskhova, which were officially approved and then printed. The employees of the state media knew very well which street poles in Leninsky Prospect had to be lit so that the cameras could capture Tereskhova, the heroine, meeting with the citizens.

Crowds and rallies were heavily controlled in the Soviet capital, especially as they celebrated the nation's space achievements. The Soviets did not want to risk empty streets, but this wasn't a problem: people flocked to see Tereskhova. Clare Boothe Luce, a former congresswoman and ambassador to Italy and Brazil, already known for her anti-communist views, was also won over. In 1963, she wrote praise for Tereskhova in *Life Magazine*, saying that Tereskhova "*orbits beyond the sexual barrier,*" and argued that this was only possible because Soviet ideology contained a message of gender equality. Evidently, feminism was stronger than anti-communism.

4.9   "No More Witches in Space"

However, Tereskhova's landing did not go according to plan: once out of the capsule, she was hungry and dehydrated, full of vomit, and had wounds and bruises caused by absolute immobility that had lasted three days in a row.

As she descended in her parachute, Valentina was hit in the face by a metal fragment, which fell as she was looking up with the visor of her helmet raised.

Coming further down, she realized that she was actually about to end up in a large lake, something she was not prepared for, but luckily, she encountered an air current that moved her and landed her on dry land.

Tereskhova returned to the ground, unconscious and with a bad bruise. When she was found by the local villagers, she accepted their peasant food, potatoes, kvass and bread, after almost three days of fasting. She also distributed her tubular space rations. Both actions were strictly against Soviet protocol.

Valentina tried to explain that it was the space food that had weakened her, but her superiors would not accept this. In doing so, she altered the results of some medical experiments on her food consumption, which Korolev strongly criticised.

"No more witches in space," he said, after Tereskhova returned to Earth.

So, Valentina's return to earth was not all honour and glory, especially as she had to deal with Air Force executives, all men unfriendly to women who usurped male positions: the officials accused her of being rebellious and incompetent. The society in those years was conservative, despite the declarations of perfect equality that we saw earlier.

Medical specialists considered both Tereskhova's adaptation to the space environment and her behaviour on the Vostok 6 mission to be unsatisfactory.

This is one of the reasons why, despite later plans in 1965 for a possible second flight, Tereskhova never returned to space. She had been envisaged as commander of a mission in which Solovyova would perform a spacewalk, but her first flight would also turn out to be her only flight.

The truth is perhaps more complicated. Not all the founding fathers of Soviet cosmonautics approved of Tereskhova's performance in space. And for this, they blamed her gender.

Throughout her flight, Tereskhova kept telling mission control that she was feeling fine, but by the third day in orbit, it became clear that she was trying to hide her fatigue.

Valentina fell asleep and missed a status call with Earth. She felt constantly nauseous, vomited, lost her appetite and was unable to carry out the planned scientific experiments. Cosmonaut Bykovsky, who could hear all communications with Earth, heard Tereskhova's calls and thought she had been crying.

Korolev was disappointed.

Surprisingly, neither she nor the five trained women in the squadron ever spoke ill of him or the way he had treated them in Star City.

Korolev had dreamed of flying into space, but would never have met the health requirements after suffering for years in Stalin's prison camps. However, he believed that, one day, his spacecraft and rockets would become so reliable and so comfortable that the health requirements would not be necessary. His comments about Tereskhova may have been dictated by frustration, because herTereskhova flight showed him the disappointing truth: any space flight would push even a young and healthy body to its limits. Let alone his ailing physique! But modern day technology has changed, and space flight gear is now different from what it was during those first flights; supposedly, the comfort level in modern day flights is higher and they do not require as strenuous a preparation as the first flights. Most importantly, modern cosmonauts deal with less severe g forces. Space tourism is on the rise, and it's now restricted only by the ticket price, rather than by health concerns or discomfort.

When Tereskhova became the first woman in space, the problems that occurred during the flight were not revealed. They were eventually discussed in the memoirs by Kamanin, Korolev and Vassili Mishin (Korolev's deputy), published after the fall of the Soviet Union, however, Tereskhova never revealed her side of the story—that is, what really happened—until 2007, when the cosmonaut was in her seventies.

Korolev was originally said to be unhappy with Tereskhova's performance, and therefore did not allow her to take manual control of the spacecraft as planned.

Vassili Mishin later stated that she was "*on the verge of psychological instability.*" Kamanin reports in his diaries that, in the government press release, an officer tried to insert a formal paragraph about Tereskhova's poor emotional state in space. The sentence spoke of

...overwhelming emotions, fatigue and a significantly reduced ability to work to complete all assigned tasks.

Kamanin disagreed: he said that the man had exaggerated in his account. After all, she only had assigned tasks for the first day. When the flight was extended for a second, and then a third day, she had virtually nothing to do.

Unfortunately, ground command did not support her during those additional days. However, it seems that Valentina did everything she could to complete the flight programme in full.

4.9.1 The Terrible Mistake

Tereskhova's version was this: the automatic orientation system of her Vostok capsule had been set up incorrectly.

She had immediately noticed one fact: as soon as she was inserted into orbit, her capsule was oriented at ninety degrees relative the desired direction. This had a terrible implication: as the braking rocket started to ignite, she would be sent to her death in a higher orbit, rather than braking for a return to Earth's atmosphere. It was later discovered that the system was misreading the trajectory data, and reversing it in the opposite direction.

She communicated the problem to ground control. But they did not believe her.

Eventually, her version was verified. But it wasn't until the second day in orbit that something was done: the right signals were sent to the spacecraft to correct the problem. It seems that it was the resolution of this problem that caused the delay in her return to Earth.

It was true that she vomited in space, and it was said that, for this reason, Korolev wanted to bring her back to Earth earlier than planned. But Tereskhova claimed that her nausea was caused by the poor quality of the space food she had been provided with. The black bread was too dry. So, she ate the few preparations that she found palatable.

She was ordered to remain tied to her seat, evidently to combat her alleged space sickness. In addition, she had wounds and haematomas caused by absolute immobility for three days in a row. On the second day, in addition to the pain in her right leg, she had an incredible itch and a sore pressure point: the helmet pressed against her shoulder.

After her landing, Sergey Korolev would approach her and beg her not to tell anyone about the orientation incident.

Looking out of the spaceship and seeing that she was moving away from Earth, Valentina had the following thoughts (which, it is not unreasonable to assume, most would have in her position):

> The blue ball that is my world is getting more and more distant. I am just an insect in the cosmos. I am losing my head! From up here, I can only make a few corrective manoeuvres, small adjustments. The course must be corrected from Earth. I am starting to swear at my three Ks! Kamanin, Korolev, Krushchev, all

three of them. Over a hundred and fifty kilometres above the Earth, there is no one to help me. All this silence! I was trained to be weightless, to accelerate to many **g**'s, I was put in the centrifuge, but no one ever trained me to be silent! And now I understand why I was chosen, a small worker, and not a pilot, not an engineer. It's fine for me to end up like Laika, I can get lost in space, nobody cares about me! My three Ks need a humble woman, who'll stay in her place! It was me, me without a father, without anyone to worry about me, I was the one! That's also why the mission was kept secret, that's why I had to lie to my mother!

The emotional shock of this situation may have been the main reason why Tereskhova was not as "productive" or was psychologically unstable. Although cosmonauts since have always been evaluated for their "psychological toughness" or ability to remain rational in the face of challenges, when faced with such a situation in reality, well, things are different and they can derail significantly. As mentioned above, as soon as she reached orbit, she realized that there was a problem with the orientation of the capsule, but the space center did not immediately accept her complaints or fix the problem. Those hours must have been agonizing and devastating.

In space, she did not keep her logbook. Tereskhova's explanation for why she didn't was that both her pencils had broken.

After being ejected from the capsule, Tereskhova was exhausted, dehydrated and hungry. She had narrowly avoided landing in the lake, thanks to the wind, but did land on the ground with considerable force. She hit her nose on her helmet, creating a dark with blue bruise. Heavy make-up was needed for the public appearances that followed. However, she feared that the make-up would conflict with her image as a worker.

4.9.2   The Lie of Re-entry

The truth is different: the doctors did have to hide Tereskhova's bruise, but only on the day after she actually landed. In fact, the entire scene was recreated the next day for filming. The cosmonaut was put back in the capsule and a few actors were hired to play the people who approached her. One of the actors opened the capsule, showing an uninjured and smiling space pioneer. Remember that the landing of a cosmonaut in a capsule was not yet within the reach of the Soviets.

After her flight, some Soviet aviation figures tried to discredit Valentina. They accused her of being drunk when she reported to the launch pad, and that she was insubordinate in orbit, ignoring direct orders from the Centre. Why? According to some sources, these people thought she should have accepted the faulty orientation of the spacecraft. That she should have preferred death to embarrassing any (male) manager in the field.

In September, after the leak of the news, it was said that Tereskhova was drunk and had created a scandal with an officer in Gorkiy. She categorically denied being drunk, but admitted to having had a confrontation with a militia captain. Kamanin defended her against the attacks, and, in the end, it was Tereskhova's opponents who were fired.

4.9.3 Conclusions

By no means was this a happy, triumphant adventure. The voyage of the first woman into space turned into an odyssey that almost ended tragically with the spacecraft shot into infinity. The return to earth was disastrous and had to be filmed again for newsreels, after a very short stay in the hospital for its star.

This trip was part of the unrelenting struggle between the two superpowers outside of the earth's atmosphere, so the Party thought it best to keep quiet, cover up the mistakes, heal the wounds and shoot another ending for a story that should never be told as it actually happened. The ending should be a smiling Valentina Tereskhova greeting the rescuers as she emerged victorious from her Vostok 6. That is how everyone remembers her.

At Korolev's request, for the next ten years, none of the witnesses were allowed to talk about what had happened, and only recently has the fact become public knowledge.

> The bread is stale. I didn't eat it. I wanted black bread, potatoes and onions. The water is fresh and pleasant. Juices and steaks are good too. I vomited once, but not because of the vestibular disorder, but because of the food. [...] We kept insisting that everything was fine and didn't talk about it. We kept it a secret for over 30 years, until the person at fault died,

explained the cosmonaut to various newspapers about the mistake that could have been fatal. Honestly, it's hard to believe her explanation. It is known that nausea is a typical problem in space flights. The severity would depend on the individual organism; she just was not lucky.

Korolev was right about Tereskhova's inadequacy, as she had insufficient technical training and had to rely on a fully automatic flight: in fact, she was unable to perform the manual re-entry manoeuvre correctly, and her colleagues on the ground had to explain to her what to do. The episode alarmed the technicians greatly, because, if the autopilot had malfunctioned, a seriously dangerous situation could have arisen.

4.9.4   The 'Red Moscow' Scent: Valentina's Image

After the Vostok-5 with Valery Bykovsky on board, the whole world was waiting for more Soviet special effects. Last time, there had been the launch of no less than two Vostoks carrying Andrijan Grigorevič Nikolaev and Pavel Popovich. No doubt, the USSR would surprise the world with something else.

The success of Valentina Tereskhova's feat was a source of great pride and propaganda value for the Soviet Union, not to mention confusion and consternation for the United States.

For one thing, the woman did not fit the stereotypes.

As historian Robert L. Griswold reveals, Americans saw the Russian working-class woman as *"ungainly, shapeless and sexless."* Many Americans imagined Soviet women as poor and unkempt, with shabby clothes and make-up, due to the miseries of communism. According to Griswold, by the late 1950s:

> ...the American conception of Soviet working-class femininity became a way of reasserting the boundaries of true femininity.

The *Chicago Tribune* dubbed Valentina Tereskhova the *"Russian blonde in space,"* even though she was not blonde. Then, there was the stereotype of the apolitical matron, played by Nina Krushcheva, Nikita Krushchev's partner. "Mrs. Krushchev" had visited the US in 1959 and was liked by virtually everyone. Although she was, in fact, *"a revolutionary in her own right,"* in the eyes of the American media, she became a kind of global grandmother, who focused only on her family and had little interest in the intrigues of the Kremlin. Griswold wrote that *"in this case, maternal ideology is more powerful than anti-communism."*

Everything changed when Tereskhova went into space. Suddenly, a new stereotype emerged on the American cultural front: professional Soviet women were *"doctors in lab coats, engineers, even a young cosmonaut in her spaceship."*

Twenty-six-year-old Tereskhova, after all, had *"travelled farther than the entire corps of American astronauts had done."* The American media had fixed, not without surprise, on her sex appeal. In the *Chicago Tribune*, it was written that Valentina wore a perfume called Red Moscow; she was compared to Ingrid Bergman.

4.9.5   Elena, the Daughter of the Stars

There are only two inexhaustible legacies that we must hope to pass on to our children: roots and wings. (Harding Carter)

It was a beautiful name for a little girl: Elena. But she was spoken of in a low voice, with the sign of the cross. There were strange rumours about her, scary gossip.

In the beginning, it was different; all of the people loved her. She was the child that all the Russians would have wanted as a daughter: the sweet little face, the big blue eyes looking up at you, on her head, a colourful scarf, like a little matryoshka doll. In the Russian shops, they sold chocolate called Aljonka, after her nickname, so that children would grow up like her: Elena Nikolaeva, daughter of the Russian cosmonaut Andrian Nikolaev, but, above all, daughter of Valentina Tereskhova, the first female cosmonaut, the first woman in space, the woman who Krushchev adored for her sweet look and steely temperament, the woman who was Krushchev's jewel, his pearl.

There were the afore-mentioned whispers that it was he who had arranged the marriage, and that the doctors wanted to see what would become of the child of two cosmonauts. Elena was born a year after her mother's return from space. She was welcomed triumphantly, like a princess, but soon dropped out of sight.

She no longer appeared in public and no pictures of her were ever published. Time passed, and nothing was known of her. But the gossip continued. Valentina's daughter must have grown up to be a monster, or the victim of an incurable star virus, so the government kept her hidden away, locked up so that no one could guess what the secret was.

There were Western spies and various tongues wagging about certain problems: the girl was deaf and mute, always under medical observation. It was even said that the Russian government had slapped the Top Secret seal on all information concerning her.

It has to be said that there was a precedent for the idea that the girl might be abnormal, and it was scary. Dogs who had been sent into space had ended up having malformed puppies. So... a mystery did exist there.

But no solutions were to be forthcoming at that time in regard to the daughter of Tereskhova. Even when Valentina divorced her husband, nothing more was heard of little Elena, as if she had never existed.

It wasn't until more than forty years later that she would quietly reappear in public. Elena, the "space baby," as she was affectionately known in the '60s. There was no mystery, and she reappeared as suddenly as she had disappeared.

Born in 1964, she had married twice, worked as a doctor in the Aeroflot health service and had had two children.

Do you see? I am alive. And I was never a monster.

She has always lived near Star City, the Russian aerospace centre. And she recounts with a smile:

When mum took me for walks in the street, people would ask me if I could see well, or if I was mute.

Mama Valentina had always been a tenacious person:

She wanted me to have a normal life, and for this very reason, she kept me hidden, saying, "We are so much better off at home."

5

From Aleksej Leonov to the Space Station

Abstract The first space walk is recounted, then the Soviet effort to reach the moon before the United States, the USSR's greatest failure, is discussed. We then look at Project Buran, which was both a success and suffered many setbacks, and end with today's collaboration between the US and Russia, as well as that between Europe and Japan.

5.1   Leonov and the First Space Walk

> How difficult it is to talk about the moon! The moon is so <u>dumb</u>. It must be the ass that the moon always makes us see… (<u>Samuel Beckett</u>)

In 2019, Alexej Leonov, the first human being to do a spacewalk, died: in 1965, he had remained outside of the capsule for twelve minutes and nine seconds. But he had risked never being able to re-enter.

Leonov was born in Listvyanka, a small town in south-eastern Russia, and at the age of 30, he took part, together with Pavel Belyayev, in a space mission on the Voskhod 2 capsule, an descendant of the Vostok that was designed to accommodate up to three cosmonauts. That mission involved the first "space-walk," and there were only two cosmonauts on board, perhaps because it was necessary to carry—and position at an exit hatch—the special inflatable "air-lock," which was indispensable for allowing the cosmonaut to exit. Once the preliminary operations had been carried out correctly, Leonov, tied to a cable,

M. R. Menzio, *The Secrets of Soviet Cosmonauts*,
https://doi.org/10.1007/978-3-031-09652-5_5

ventured out. Leaving the capsule, he operated a video camera to film and broadcast the scene live.

In his book, *Two Sides of the Moon*, written together with the American astronaut David Scott—the seventh man to set foot on the surface of the Moon—Leonov writes that, during those minutes, he felt like

...a seagull with outstretched wings, soaring above the Earth.

For twelve long minutes, attached to a cable, he "walked" in space. Two months later, he told the American periodical *Life Magazine*:

I was fully concentrated, cold-blooded and relatively unexcited. But the sight was extraordinary: the stars were not shining, everything was still, except the earth.

What he did not reveal was that he and Belyayev were lucky to have survived. They had defied death.

Several teams, both Russian and American, had already orbited the Earth, so Voskhod 2 was to push the boundaries of spaceflight a step further by including a spacewalk, or EVA, in the mission.

What happened that was so tragic?

About ninety minutes after take-off, in order to get out of the capsule for his first walk, Leonov walked to the ship's inflatable decompression chamber, tied a five-metre cable around his chest, opened the hatch and ventured out into the unknown, with only his suit to protect him. It was the first time a human being had left the safety of a spaceship to float in orbit without footholds. A camera, which Leonov had activated on his way out, captured the extraordinary moment. A second camera attached to Leonov's chest was less successful, because Leonov's suit unexpectedly inflated in response to the change in atmospheric pressure. Because of this, he could not reach the shutter button on his thigh. Leonov apparently didn't care much for his space suit, and, instead, was entranced by the perfect view of Earth.

He writes:

I heard Pasha telling me: "It's time to come back inside." I had already been floating in space for ten minutes. At that moment, my mind went back to childhood for a second, to my mother opening a window in the house and calling me, while I was playing in the yard with my friends: "Lyosha, it's time to come back inside." Reluctantly, I accepted that I had to get back into the ship again.

But getting back on Voskhod 2 wasn't going to be so easy. Leonov's inflated suit had become stiff over the course of the twelve-minute spacewalk, and was too large to allow him to re-enter through the hatch. The cosmonaut risked being left outside. Realising that Leonov was stranded in space, Soviet mission control interrupted the live telecast and replaced it with a rendition of Mozart's Requiem. Meanwhile, the clock was ticking.

I knew not to panic,

he said later. The cosmonaut opened the oxygen valve, which deflated the suit, but also meant that he risked running out of air. The manoeuvre proved successful, and his incredible calmness saved the day. Fortunately, he managed to squeeze himself into the module before running out of air, and before being overcome by heat stroke and decompression sickness.

When I entered the capsule, I was drenched in sweat and my heart was racing. But that was only the beginning of the problems.

On board, the oxygen pressure had risen to warning levels, and one spark could set everything off. The level subsequently returned to normal, but without the two astronauts knowing how it had done so. After unloading the decompression chamber and preparing for re-entry, Leonov noticed that the automatic guidance system was not working. With only a few minutes to spare before they reached the atmosphere, Leonov and Belyayev had to figure out very quickly how to manually orient themselves and calculate an ignition sequence for the retro-engines that would allow them to re-enter safely within Soviet borders. They succeeded—sort of. After an extremely steep descent at very high pressure, Voskhod 2 landed in Soviet territory, in a wolf-infested Siberian forest, near the town of Solikamsk, well off course, some four hundred kilometres from the planned landing site. Leonov remembers Beyayev asking him when he thought help would arrive. And he joked:

In three months, maybe, they will find us with sleds with dogs.

But as luck would have it, the duo didn't have to endure the freezing temperatures for too long. As the Voskhod landed with the cosmonauts on board, the two spent a night hiding in the capsule. But they were without heating, apart from a few blankets that a helicopter, unable to descend because of the bad weather, had managed to parachute down to them. Recovering this cargo, however, meant dealing with wolves and bears. Subsequent missions were said to include a TP82 pistol among the equipment on board.

Fortunately, the following day, the two were found by a rescue team. The crew spent a second night in a shelter with a fireplace before arriving at safe harbours and world fame. Unfortunately, Belyayev was only able to enjoy it for five years, dying of peritonitis in 1970. Leonov, on the other hand, lived a long life, achieved a major milestone in space travel and toured the world as a living testimonial to the Russian space program.

Once the full story was told, his dramatic experience added interest to the events. In his own words:

> I keep thinking back to the mission and realise mistakes we could have avoided. It could have been a tragedy. Everything was on the edge.

The mission was saved and, with it, the Soviet record in space. The first US astronaut to recreate the "walk" was Ed White, but not until three months later. Too long a period for it not to go down in the history books as another setback for the stars and stripes in their conquest of space.

This was not Leonov's only time in space. He went back in 1975, as part of the Soyuz-Apollo programme, the first in which the United States and the Soviet Union cooperated in linking the Russian Soyuz to the US Apollo capsule.

Leonov earned the title of the "first astronaut" to move far from Earth, into the "deepest blackness," in the race to conquer space that engaged the US and the USSR for decades.

The eleventh Soviet cosmonaut, Leonov was considered a legend of space exploration in Russia, but was also celebrated by the Americans. As a young man, he showed an interest in art and drawing, and wanted to attend the Riga Art Academy. This enabled him, in the last years of his life, to *paint pictures* of the highlights of his career.

But the issue of the inflated suit was not the only risk Leonov ran in his life. On January 22, 1969, in front of the Kremlin walls, an armed terrorist shot at the motorcade in which Leonid Brezhnev was travelling, killing one person and wounding four others. Fate, however, worked in Brezhnev's favour, because the car carrying the cosmonauts was mistaken for his car. The terrorist fired ten shots, killing the driver and wounding Andriana Nikolaeva and Georgij Beregovoy. Aleksej Leonov and Valentina Tereskhova were also travelling in the same car.

On October 11, 2019, Alexei Leonov passed away, aged 85, after a long illness. The eleventh Soviet cosmonaut, he was considered a legend of space exploration.

With his death, some indiscretions, published in an article in *Air & Space* by Anatoly Zak, a journalist specialising in the history of space exploration, have come to light. Zak claims that his piece refers to documents that have only recently been made public: more extensive footage, the original logbook, reports signed personally by Leonov immediately after the mission. According to Zak, there is no account in this documentation of the heroic act that has been told to the world for decades.

Instead, it is likely that the account of what happened only emphasised a situation that the Russian space experts had feared might happen, at least in part, and for which a way out had been planned and executed. In essence, it was foreseen that the lack of external pressure would cause the suit to inflate, and so tests were conducted on how to vent it by letting the oxygen escape. Of course, the extent of the discomfort inflicted upon Leonov by the inflated suit, the stiffening of the fabric of which it was made and that prevented the cosmonaut's movements, the impossibility of operating the video camera that he was wearing, are all facts. And they seem to confirm that, in spite of the possible evidence, there was a real risk of tragedy.

5.2 Hercules N-1

The N-1 rocket (or Hercules N-1) was a Soviet multistage launcher designed to take cosmonauts to the moon. While the US pooled all funding for a single programme, namely, Apollo, the USSR fragmented its resources into two design groups and nine different types of launcher. But it was Korolev's group that designed the N-1 rocket. The most complex part of the programme was the choice of fuel. Korolev's rival, Valentin Petrovich Gluško, was thinking of a rocket with only one combustion chamber, whereas large carrier rockets usually had four. And for fuel, he wanted a mixture of paraffin and oxygen, and not UDMH (asymmetric dilmetrazine).

The N-1 rocket was 105 m tall and was supposed to be able to carry 70 tonnes into orbit. But it was too big, and had constant technical problems. Additionally, each rocket had to have another 30 Kuznecov rockets of its own, 24 in an outer ring and 6 in an inner ring. The air passing between the two concentric circles increased the rocket's performance, boosting its efficiency and reducing paraffin consumption.

Let's look at some details. N-1 was divided into five stages; it had a maximum diameter of 17 m and a take-off mass of 2,735 tonnes. It was intended to carry 3 cosmonauts to the Moon and back to Earth. The 5 blocks, from the ground upwards, were named A, B, W, G and D. Cone block A was the first stage. It was 30 m tall, 17 m in diameter at the base and 10 m at the head. Fully loaded, it had a mass of 1,870 tonnes, accounting for two-thirds of the total mass. It weighed so much because the stored fuel could amount to 1,690 tonnes of liquid oxygen and RP-1 (refined paraffin). Thrust was 45,000 KN and the fuel was burned in just 113 s. When Block A had run out of fuel, it was released and the second block, B, was launched, although no launch ever reached this stage. B was similar to A: 20.5 m tall, a base diameter like that at the top of A and ending with a width of 6.8 m. It had eight NK-15 engines modified for high altitudes, arranged in a single ring, giving a thrust of 14,000 KN and burning 500 tonnes of fuel in 120 s. If B had dropped out, W would have gone into operation, working outside of the atmosphere, where the thrust requirements are lower. It was 11 m tall, had a maximum diameter of 5.5 m and a thrust of 1600 KN for 375 s, the time for possible manoeuvres. On the other hand, the fourth block, G, was not really part of the N-1, as it had to carry cargo from low Earth orbit to the Moon. It was a cylinder, 9 m × 4.1 m. It had a single engine, a capacity of 55 tons and 440 s of fuel life. D, the only block certified for human flight, cylindrical in shape, 5.7 m tall and 2.9 m in diameter, was the braking level for lunar orbit—so that the lander could descend—as well as the level for transferring the crew to Earth. The engine was re-ignitable, ran on liquid oxygen and RP-1, gave a thrust of 83 KN, and had 14.5 tonnes of fuel burning in 600 s.

Compared to a production of 11 or 12 launchers, only four were launched. Whereas, in the US, all engines in a batch were tested, in the USSR, one engine was tested on a random basis. If it was unsafe, the whole batch was discarded. But if one of them worked, the whole batch was accepted! Even though there was no guarantee that the other engines would work too! It was only from 1970 on that some engines were allowed to undergo ignition tests before launch.

Korolev's death led both to the splitting of the lunar programme into two and to serious delays. Vasily Pavlovič Mišin, one of the builders of the Semyorka R-7 rocket, was appointed as project manager. Work was accelerated and construction of the first N-1 rockets began. Due to haste, the functioning of the first stage of the rocket was not verified and the assembly stage was moved to Baikonur.

Four launch attempts were made:

– February 21, 1969: A fire broke out in the engine section, and just as quickly went out.
– July 1969, three weeks before Apollo 11: There was a technical failure of one of the propellant pumps, resulting in an explosion and disintegration.

At that point, all N-1 rockets were fitted with a system that made them take off at a slight angle, so that the rocket would not fall back on its head if it failed.

- June 26, 1971: The rocket lifted off, but began to wobble, and 50 s later, control was lost.
- November 23, 1972: At 9:11:52 a.m., from launch pad 110 of the Baikonur Cosmodrome, the giant N-1 lifted off. On board was a LOK lunar module and all of the Soviets' now slim hopes for "their own" conquest of the Moon. The rocket lifted off, but after 107 s, all hope was lost. Shortly before the first stage lifted off, the launcher was destroyed. Apparently, the reason was the explosion of a liquid oxygen pump, due to the excessive pressure in the tubes generated by an early shutdown of the 6 internal engines of the first stage.
- At that point, the programme was cancelled, and the story of the four launch attempts was kept secret for years.

The subject of the "Russian Moon" was addressed in great detail by engineer Boris Chertok. In 1946, Chertok began working in the design office for spacecraft, eventually becoming deputy chief designer of the Soviet space programme and retiring in 1992, at the age of eighty. Born in 1912 under the Russian Empire, he saw the 1917 revolution, the First and Second World Wars, and the fall of the Berlin Wall, passing away in 2011, having almost reached the 100-year mark. He received 32 honours and awards, including 'Hero of Socialist Labour.'

The most comprehensive history of Soviet attempts to go to the Moon can be credited to him, in the fourth volume of his monumental book *Rockets and People*.

Here are some extracts from the introduction written by his editor, Asif A. Siddiqi:

In the summer of 1989, Soviet censors finally allowed journalists to write about an episode of Soviet history that had officially never happened: the massive Soviet effort to compete with Apollo in the 1960s to land a human being on the Moon.
[...]
   As more and more information emerged in the early 1990s, some salient features began to emerge: that the program had been massive, that it had involved the development of a super booster known as the N-1, that all efforts to beat the Americans had failed, and that evidence of the program had been whitewashed out of existence.
[...]
   It has become increasingly clear to historians that it would be impossible to understand the early history of the Soviet space program without accounting for

the motivations and operations of the human lunar landing program. By the late 1960s, the N1-L3 project constituted about 20 percent of annual budget expenditures on Soviet space exploration; by some estimates, total spending on the Moon program may have been about 4–4.5 billion rubles, which roughly translated to about 12–13.5 billion dollars in early 1970s numbers.

But beyond the numbers, the program was undoubtedly one of the most dramatic episodes in the history of the Soviet space program. During the eventful and troubled period that Chertok covers in this volume, from about 1968 to 1974, the Korolev design bureau, now led by the talented but flawed Vasilij Mišin, stumbled from one setback to another. The heart of the program during these years was the giant N-1 rocket, a massive and continually evolving technological system whose development was hobbled by difficult compromises in technical approaches, fighting between leading chief designers, lack of money, and an absence of commitment from the Soviet military, the primary operator of Soviet space infrastructure.

Chertok begins his narrative with a discussion of the origins of the N-1 in the early 1960s and the acrimonious disagreement between Sergey Korolev, the chief designer of spacecraft and launch vehicles, and Valentin Glushko, the chief designer of liquid-propellant rocket engines. On one level, theirs was a disagreement over arcane technical issues, particularly over the choice of propellants for the N-1, but at a deeper level, the dispute involved fundamental differences over the future of the Soviet space program. Korolev and Glushko's differences over propellants date back to the 1930s when Glushko had embraced storable, hypergolic, and toxic propellants for his innovative engines. By the 1940s, Korolev, meanwhile, had begun to favor cryogenic propellants and believed that a particular cryogenic combination, liquid hydrogen and liquid oxygen, was the most efficient way forward. Korolev was not alone in this belief. In the United States, NASA had invested significant amounts in developing such engines, but Glushko had an important ally on his side, the military. When Korolev and Glushko refused to come to an agreement, a third party, Nikolay Kuznetsov's design bureau in the city of Kuybyshev (now Samara), was tasked with the critical assignment to develop the engines of the N-1.

Having known both Korolev and Glushko, Chertok has much to say about the relationship between the two giants of the Soviet space program. He shows that they enjoyed a collegial and friendly rapport well into the 1950s. He reproduces a congratulatory telegram from Korolev to Glushko upon the latter's election as a corresponding member of the Academy of Sciences. It obviously reflects a warmth and respect in their relationship that completely disappeared by the early 1960s as the N-1 program ground down in rancorous meetings and angry memos. [...]

One of the main challenges of developing the N-1's engines was the decision to forego integrated ground testing of the first stage, a critical lapse in judgment that could have saved the engineers from the many launch accidents.

Chertok's descriptions of the four launches of the N-1 (two in 1969, one in 1971, and one in 1972) are superb. He delves into great technical detail but also brings into relief all the human emotions of the thousands of engineers, managers, and servicemen and -women involved in these massive undertakings. His accounts *are particularly valuable for giving details of the process of investigations into the disasters, thus providing a unique perspective into how the technical frequently intersected with the political and the personal. His account of the investigation into the last N-1 failure in 1972 confirms that the process was fractured by factional politics, one side representing the makers of the rocket (the Mishin design bureau) and other representing the engine makers (the Kuznetsov design bureau). Some from the former, such as Vasiliy Mishin, made the critical error of allying themselves with the latter, which contributed to their downfall. Historians have plenty of examples of the impossibility of separating out such technological, political, and personal factors in the function of large-scale technological systems, but Chertok's descriptions give a previously unseen perspective into the operation of Soviet "Big Science."*

Chertok devotes a lengthy portion of the manuscript (five chapters!) to the emergence of the piloted space station program from 1969 to 1971. We see how the station program, later called Salyut, was essentially a "rebel" movement within the Mishin design bureau to salvage something substantive in the aftermath of two failed launches of the N-1. These "rebels," who included Chertok himself, were able to appropriate hardware originally developed for a military space station program known as Almaz—developed by the design bureau of Vladimir Chelomey—and use it as a foundation to develop a "quick" civilian space station. This act effectively redirected resources from the faltering human lunar program into a new stream of work—piloted Earth orbital stations—that became the mainstay of the Soviet (and later Russian) space program for the next 40 years. The station that Mishin's engineers designed and launched—the so-called Long-Duration Orbital Station (DOS)—became the basis for the series of Salyut stations launched in the 1970s and 1980s, the core of the Mir space station launched in 1986, and eventually the Zvezda core of the International Space Station (ISS). [...]

Chertok's account of the dramatic mission of Soyuz-11 in the summer of 1971 is particularly moving. The flight began with an episode that would haunt the living: in the days leading up the launch, the primary crew of Aleksey Leonov, Valeriy Kubasov, and Petr Kolodin were replaced by the backup crew of Georgiy Dobrovolskiy, Vladislav Volkov, and Viktor Patsayev when Kubasov apparently developed a problem in his lungs. The original backup crew flew the mission and dealt with some taxing challenges such as a fire on board the station and personality conflicts, and then they were tragically killed on reentry when the pressurized atmosphere of the Soyuz spacecraft was sucked out due to an unexpected leak. The funeral of these three cosmonauts was made all the more painful for, only days before, Chertok had lost one of his closest lifelong friends, the engine chief designer Aleksey Isayev.

A chapter near the end of the manuscript is devoted to the cataclysmic changes in the management of the Soviet space program that took place in 1974: Mishin was fired from his post, the giant Korolev and Glushko organizations were combined into a single entity known as NPO Energiya, and Glushko was put in charge. These changes also coincided with the suspension of the N-1 program and the beginning of what would evolve in later years into the Energiya-Buran reusable space trans-portation system, another enormously expensive endeavor that would yield very little for the Soviet space program. [...]

A recent collection of primary source documents on Glushko's engineering work suggests that Glushko came to the table with incredibly ambitious plans to replace the N-1 and that these plans had to be downsized significantly by the time that the final decree on the system was issued in February 1976. [...]

The breadth of Chertok's recollections, covering nearly 100 years, makes it unique. In the absence of any syncretic work by a professional historian in the Russian language on the history of the Soviet space program, the contents of "Rockets and People" represent the most dominant narrative available.

This was how it began, as Boris Chertok writes:

"After intensive negotiations, meetings, and consultations, President Kennedy reached a decision, and on 25 May 1961, he addressed Congress and in fact all Americans. In his speech he said: "Now it is time to take longer strides, time for a great new American enterprise, time for this nation to take a clearly leading role in space achievement, which in many ways may hold the key to our future on Earth.... I believe that this nation should commit itself to achieving the goal, before this decade is out, of landing a man on the Moon and returning him safely to the Earth. No single space project in this period will be more impressive to mankind, or more important for the long-range exploration of space; and none will be so difficult or expensive to accomplish." [...]

"Congress approved the decision to send an American to the Moon almost unan-imously. The mass media showed broad support. Soon after Kennedy's address, Keldysh paid a visit to Korolev at OKB-1 to discuss our comparable program. He said that Krushchev had asked him: 'How serious is President Kennedy's announce-ment about landing a man on the Moon?' 'I told Nikita Sergeyevich,' said Keldysh, 'that technically the mission can be accomplished, but it will require a very large amount of resources. They will have to be found at the expense of other programs.' Nikita Sergeyevich was obviously worried and said that at that time we were the indisputable leaders in world cosmonautics. However, the U.S. had already passed us in the lunar program because right away it was proclaimed a national cause:— For all of us must work to put him [the first man on the Moon] there-'Space dollars' had begun to penetrate into almost every area of the American economy. Thus, the entire American public was in control of the preparations for a landing on the Moon. Unlike the Soviet space projects, the U.S. lunar program was not classified. The U.S. mass media did not cover up the fact that Yuri Gagarin's flight on 12 April 1961 had shocked the nation. Americans were afraid that the rocket that had carried Gagarin was capable of delivering an enormously powerful hydrogen bomb to any point on the globe. One has to give President Kennedy credit. He was able to find a quick response, reassure a nation, and simultaneously mobilize it to achieve

great feats. Titov's flight on 6 August 1961 struck another blow against American public opinion. But the lunar "psychotherapy" had already begun to take effect. On 12 September 1962, speaking at Rice University, Kennedy declared:

We choose to go to the moon in this decade and do the other things, not because they are easy, but because they are hard, because that goal will serve to organize and measure the best of our energies and skills, because that challenge is one that we are willing to accept, one we are unwilling to postpone, and one which we intend to win, and the others, too. [...]

"Reliability and safety were the strict conditions for all phases of the U.S. lunar program. They were achieved as a result of thorough ground-based developmental testing so that the only optimization performed in flight was what couldn't be performed on the ground given the level of technology at that time. The Americans succeeded in achieving these results thanks to the creation of a large experimental facility for performing ground tests on each stage of the rocket and all the modules of the lunar vehicle. It is much easier to take measurements during ground-based testing. Their accuracy is increased, and it is possible to analyze them thoroughly after the tests. The very high costs of flight-testing also dictated this principle of the maximum use of ground-based experimental development. The Americans made it their goal to reduce flight-testing to a minimum. Our scrimping on ground-based development testing confirmed the old saying that if you buy cheaply, you pay dearly. The Americans spared no expense on ground-based developmental tests and conducted them on an unprecedented scale." [...]

"Each series-produced engine underwent standard firing tests at least three times before flight: twice before delivery and a third time as part of the corresponding rocket stage. One must keep in mind that to achieve the required reliability, both we and the Americans had two basic types of tests." [...]

"The transfer to von Braun's German team of all technical management over launch-vehicle production in its entirety played a decisive role in the success of the American lunar program." [...]

"A comparison of the status of operations on the respective lunar programs in the U.S. and USSR in early 1964 shows that we were at least two years behind on the project as a whole."

"On 13 May 1946, the first decree calling for the organization of operations for the production of long-range ballistic missiles in general and the R-1 missile in particular was issued. Fifteen years later, on 13 May 1961, the order went out to produce the N-1 rocket in 1965. We were actually quite serious that we would produce it in 1965! Perhaps not for a landing expedition to the Moon, but certainly for defense and other purposes. Overconfidently, we sought to produce the desired rather than the feasible. Of course, the authorities encouraged us to behave like that." [...]

"Korolev's will and even Central Committee and government decrees proved insufficient for the oxygen-hydrogen liquid-propellant rocket engines under development for the N1-L3 program to be produced in time to take their place on the <lunar> rocket." […]

In May 1962, N-1 was assigned the following primary objectives:

a. *Insert heavy space vehicles (KLA) into orbit around Earth to study the nature of cosmic radiation, the origins and development of the planets, solar radiation, the nature of gravity, and the physical conditions on the nearest planets, and to discover organic life-forms under conditions different from those on Earth, etc.*
b. *The insertion of automatic and piloted heavy satellites into high orbits to relay television and radio broadcasts, for weather forecasting, etc.*
c. *When necessary, the insertion of heavy automatic and piloted military stations capable of staying in orbit for long periods of time and making it possible to perform a maneuver for the simultaneous orbital insertion of a large number of military satellites.*

And then, obviously, there was the Moon.

"The plan declared the main phases for the further exploration of space:

- *Execute circumlunar flight of a spacecraft with a crew of two or three cosmonauts;*
- *Insert a spacecraft into lunar orbit, land on the Moon, explore its surface, and return to Earth;*
- *Conduct an expedition to the lunar surface to study the soil and topography and to search for a site for a research facility on the Moon;*
- *Build a research facility on the Moon and set up transport systems between Earth and the Moon;*
- *Conduct a flight with a crew of two or three cosmonauts around Mars and Venus and return to Earth;*
- *Conduct expeditions to the surface of Mars and Venus and select sites for research facilities;*
- *Build research facilities on Mars and set up transport systems between Earth and other planets; and*
- *Launch automatic spacecraft to explore circumsolar space and the distant planets of the solar system (Jupiter, Saturn, etc.).*

Even 45 years later, the text cited above seems like an amazing cascade of missions capable of captivating thousands of enthusiasts. It is unfortunate that not only were none of these missions ever announced to the public, or even to the scientific community, but they were also shrouded under a "top secret" classification. One might ask us, 'In 1962, did you really not understand that, aside from a lunar landing and the dispatching of automatic stations, the remaining phases should have been planned for the 21st century?'" [...]

Once again, an exaggerated future, as Ponomareva had remarked about spaceflight. Chertok goes on:

"In July 1962, the expert commission approved our draft plan for the N-1 launch vehicle capable of inserting a satellite with a payload mass of 75 tons into circular orbit at an altitude of 300 kilometers. Academy of Sciences President M. V. Keldysh approved the findings of the expert commission on the N-1 project, which named defense rather than lunar missions as the primary tasks for the N-1."

After many discussions and fights, practical problems and bureaucratic issues...

"Theoretically, the three-launch configuration would enable us to compensate for the large number of advantages of the American design, which used hydrogen fuel for the second and third stages of the Saturn V launch vehicle. Of course, in terms of cost-effectiveness and general system reliability at that time, we were losing." [...]

"And really, how is a highly placed official supposed to react to complaints about insufficient funding for a program involving a lunar expedition in the distant future, if this very pushy chief designer has had four failures in a row during launches of automatic stations to Venus and for the soft landing of automatic vehicles on the Moon on 21 March, 27 March, 2 April, and 20 April?" [...]

"Under these circumstances we need to reconsider the concept of the three-launch profile with a landing on the Moon. The whole time they will accuse us of having a complicated, unreliable, and expensive version compared with the Americans' single-launch profile. But the Americans already have a hydrogen engine and it's already flying, while all our engine specialists have [....] only promises,' concluded Korolev." [...]

"After Kalmykov had pulled out of me an approximate list of problems that needed to be solved, he asked: 'Tell me frankly, forgetting for a minute that I am a minister, a member of the Central Committee and all that—you want to do all of this in three years so that in 1967, the 50th anniversary of the Revolution, you can have a fully tested system, and on 7 November, after returning from the Moon, our cosmonauts can stand on Lenin's Mausoleum [and watch the parade go by]? Is this really what you thought?' I confessed that I wasn't certain that this date was realistic, but if a later date were proposed, we would risk having the project prolonged indefinitely. 'This is not a reason'—objected Kalmykov—'I believe that everyone needs not three years, but six or seven years. Considering the actual work

load on the industry, you all deserve to have monuments erected to you in your lifetime if our cosmonauts fly to the Moon and return safely before 1970." [...]

"The calculations specialists showed that to ensure absolute superiority over the U.S., a 200-ton rocket complex should be assembled in Earth orbit using three N-1 rockets. This would require three N-1 rockets or 20 UR-500 rockets. In this case, we could manage a lunar landing of a vehicle weighing 21 metric tons and return a vehicle weighing 5 tons to Earth. All the economic calculations were in favor of the N-1. Despite the positive assessment of the leading institute, Korolev firmly decided to move forward only with the single-launch format." [...]

"Krushchev was known to be an emotional and unpredictable man, and also his approach to the moon landing program made no exception. In 1964, during a meeting with Sergei Korolev, Krushchev told him that the money for that program was running out. Then, in 1965, Korolev and Keldysh, with Ustinov's support, approached him again: 'Are we going to fly to the Moon or not?' These were Krushchev's instructions: 'Don't let the Americans have the Moon! Whatever resources you need, we'll find them.'"

The topic (Soviets on the Moon) also aroused a lot of interest in Europe, from journalists and writers. In his book *Luna Rossa*, astrophysicist Massimo Capaccioli shows how the Soviets were indeed the first to circumnavigate the Moon, photographing its hidden face and touching the ground with a robot. But they suffered a crushing defeat at the hands of NASA. How, then, was such a setback possible, given the great superiority of Soviet space missions until the mid-1960s? Laika, Gagarin, Tereskhova, Leonov... Capaccioli explains it as follows:

"The Americans won, because they could not lose. The delay accumulated by the Soviets was catastrophically damaging from all points of view, something that they realised after the enterprise of Gagarin, and it was absolutely necessary to recover."

Moreover, the US staked everything on one card: the Moon; they did not waste energy working in parallel on both the Moon landing and building a space station around the Earth, and they invested incredible amounts of money.

After Kennedy's appeal, NASA's budget grew to $5 billion a year, while the Soviets, who had only a tenth of that amount at their disposal, saw Sergey Korolev, the best Soviet rocket designer, the 'glavniy konstruktor' (chief builder) of the Soviet Union's space programme, die. But, unfortunately for the Soviets, the N-1's rival was Saturn V, designed by Korolev's US rival: Wernher von Braun.

It should be noted that, until 1961, the US had had no inkling of the possible USSR conquests, but after Gagarin's adventure, they woke up.

...to the sound of the International,

as Capaccioli says, and they were lucky, because:

> *"The Russians came up with a faulty strategy regarding the motorisation of their missiles: for the transport of a capsule with crew on board, they thought of the N-1 rocket, with 30 engines! Too many to control all at once."*

Everyone was aware that setting foot on the Moon meant unimaginable success in the eyes of foreigners.

> *"Khrushchev was sitting on a pin, because of the Hungarian uprising, the Suez Canal crisis and the criticism of Stalin, which strengthened and weakened him at the same time."*

On the other hand, Kennedy was

> ...a much-discussed Catholic president. He had to deal immediately with the failed Bay of Pigs invasion and had inherited a series of global conflicts in which he had no primary interest. He was a president who had to build an image for himself, and the Moon offered him the opportunity to write himself into history.

The failure of the Soviet lunar programme is widely attributed to the division of it into two rival sectors: that of launchers for flight, commanded by Korolev, and that for landing on the Moon, commanded by Chelomej.

As Aleksej Leonov pointed out in *Komsomolskaja Pravda* in 2010:

> The very complicated relations between Korolev and Chelomej, and their rivalry, have damaged our common cause.

Boris Chertok himself speaks of the rivalry between the two designers, the consequent reduction of Korolev's N-1 project to a minimum, and budget cuts. All of this led to failure.

Moreover, after the fall of Khrushchev, the central people in the programme were ousted in 1964. When Korolev died, Leonov said that:

> For us astronauts, it was almost the end of the world.

According to him, once Korolev was gone, the programme was no longer among the government's priorities. What mattered were nuclear weapons.

Thus, the USSR lagged further and further behind the US in space technology. Moreover, US rockets were powered by liquid hydrogen, which was more energy-efficient than the Soviet kerosene-based fuel.

There is nothing left of the N-1s: all the spacecrafts under construction have been dismantled. Some parts were even used to build pig pens. But one disobedient employee stored some engines. Twenty years later, one of these was taken to the US, and its excellent performance was confirmed. Of the 150 engines shelved, Russia sold 36 to Aerojet at the end of the 1990s, which also bought the licence to produce more, renaming them AJ26-58, AJ26-59 and AJ26-62. All of this is further proof of the very high level of technology of Soviet liquid propellant engines.

5.3   Luna 15, the Secret Moon of the Russians

Have you ever been to the Moon? Aaah, it is a very interesting place, the King and Queen are delightful! (Rudolf Erich Raspe, Baron Munchausen)

Russia Beyond[1] tells us about Luna 15, the USSR's secret programme to land on the Moon before Apollo 11. The probe had no humans on board and, at that point, its only objective was to land on the Moon, collect soil samples to bring back to Earth, return and, of course, do so before the Americans.

So, within the exact same minutes that all nations were holding their breath following Neil Armstrong's historic landing on the Moon, the Soviet secret mission was also aiming for the Moon. Attempting to reach our satellite at all costs, the Soviets wanted to win that chapter of the space race at the last minute.

The Moon programme appeared in 1958, well before the US Apollo, and involved sending three spaceshifts to our satellite. Over the years, the Soviets, who had already won the race for the first satellite in orbit, would go on to send the first animal, the first man, and the first woman, as well as achieving the first space walk with Leonov, confirming their supremacy in space, which the whole world believed in at the time.

On the fourth attempt, the Soviet Union launched the Luna 1 station into the air, the first spacecraft to leave Earth's orbit (although it did pass the Moon). In 1959, Luna 3 took the first photos of the dark side of the Moon's surface.

---

[1] https://it.rbth.com/storia/83238-luna-15-il-programma-segreto, *di Ekaterina Sinelshchikova, 30-10-2019.*

As its name indicates, Luna 15 was the 15th officially announced mission (in practice, it was the 31st). Many of the probes failed to reach Earth orbit, others did not even get off the ground. The Soviet government preferred to cover up the failures, knowing that there was still a lot of work to be done.

When the date of July 16, 1969, was made public—the departure of the Americans on Apollo 11 to set foot on the Moon—the Soviets could not wait any longer. They were certainly not in a position to send cosmonauts to the Moon, but there might be another prize, a way to sweeten the bitterness left by the knowledge that they had lost that leg of the space race. The mission objectives could only be limited. Of course, it was a secret mission, at least in content. The scheduled launch date, July 13, 1969, was designed to allow the craft to return to Earth a few days before the Americans.

Naturally, the announcements by the US and the USSR of two almost simultaneous missions (June 13 for Luna 15 and June 16 for Apollo 11), both heading for the Moon, aroused numerous perplexities and questions within the scientific community. It was thought to be bizarre: two spaceshifts simultaneously transmitting radio signals from the Moon to Earth! And Luna 15 was (or was supposed to be) a secret. NASA feared interference, and thus sent the Apollo 8 commander, Frank Borman, to the USSR. He was on good terms with the Soviets and was the first American astronaut to visit the country. He returned home confirming that there would be no problems, as the transmission channels would not interfere.

In July 1969, three days before Apollo 11, the USSR shuttle got "stuck" in lunar orbit, possibly due to a malfunction or to the Moon's gravitational field, the full effects of which had not been foreseen. The fact of the matter is that Luna 15 travelled 52 revolutions around the Moon while Soviet physicists were desperately trying to find a solution to the landing problem.

In the meantime, Apollo 11 had arrived in lunar orbit, the LEM had detached itself from the Service Module (CSM) and touched down on the Moon and everything was going according to plan.

At the Jodrell Bank Observatory, British scholars followed the stages of the tragicomic spectacle, listening to the words of both the US victory and the USSR defeat, with a recording that the world only came to know in 2009, forty years later. Scholars realised that Luna 15 was not just there to photograph the Moon's soil, it wanted to land! They then said:

It's landing! This is really a drama of the first order.

This happened on July 21, two days after the Apollo 11 moon landing; the spacecraft subsequently started its engines and began its descent. Unfortunately, transmissions ceased 4 min after the manoeuvre began, at an altitude of about 3 km. Luna 15 smashed into a lunar mountain. And stayed there.

However, it was in connection with the Luna 15 mission that the first case of space cooperation between the US and the USSR occurred. In fact, after the announcement of the launch of the Soviet spacecraft, there were fears at NASA that the routes might coincide, and so Frank Borman, just returned from his visit to the Soviet Union, contacted the head of the Soviet Academy of Sciences Mstislav V. Keldysh, expressing these concerns. A few hours later, in a rare occurrence for Soviet authorities, Keldysh transmitted his flight plan so as to ensure that it would not collide with Apollo 11.

If there was still any doubt about the US lunar landing, the fact that the Soviets felt it was a defeat shows that the Americans really did go to the Moon.

Later, in his book *Challenge to Apollo*, space historian Asif Siddiqi wrote:

> The whole mission was shrouded in a veil of irony: even if there had not been a critical 18-hour delay in the landing attempt, and even if Luna 15 had succeeded in landing, collecting soil samples and returning safely to Earth, its small capsule would have returned to Soviet territory two hours and four minutes after the Apollo 11 landing. The space race was over before it had even begun.

The failure of Luna 15 is related to those of the N-1 launcher: in fact, of the four (all failed) N-1 Hercules launch attempts, two occurred in 1969, the first on February 21 and the second on July 3. Incidentally, it should be noted that, in all the failed launch attempts, N-1 was carrying an experimental lunar module.

This shows how the Soviet dispersion of resources to different missions, sometimes with common objectives but with different means, did not "pay off."

In 1970, the Soviets tried to regain some of their prestige by sending to the Moon the Luna 16 probe, which brought back lunar soil samples to Earth. Another successful Soviet programme was Lunochod, which managed to land two remote-controlled rovers on the lunar surface.

5.4   Project Buran: An Expensive Exercise in Style

In February 1976, the perceived military threat of the US Space Shuttle led the USSR to the decision to build a Soviet equivalent. The project continued until 1979, when it was frozen. Although Energiya-Buran was mainly a programme of broken promises and shattered dreams, it represented a major technological breakthrough for the Soviet Union, even though the rocket engine launch tests often ran into problems.

The thing that stands out about the design of the Buran (literally 'blizzard') is the high level of ambition that the Soviet Union had in the space race. For the Russian government, the American project was a threat: it could be a planetary bomber, and they needed to take action. The project officially began on February 12, 1976, but the Soviets were already two years behind the Americans. In fact, they had resoundingly lost the "battle for the Moon" after spending a lot of money on the SATURN V competitor, the N-1 launcher, that mastodon with the unpleasant tendency to explode (only four launch attempts, all failed). The Americans were going back and forth from the Moon, they had created an aerospace industry and—according to the Soviet military spheres—it had to be used for something: perhaps they were thinking of a future of space stations and had built the Shuttle as a kind of space tram.

And what did the Russians do? They argued and, in the end, decided not to waste any time: the choice settled on a spaceplane that was a little bit, just a little bit, based on the American version. It is thought that the spacecraft was made by stealing or copying public details of the design and adapting them to Soviet technology: the similarity of the wing profiles of the two shuttles is striking. But the Soviets also thought that they could kill two birds with one stone by building a shuttle that was better than the Shuttle, and to save a few rubles, they designed and built it with military use in mind.

The story of the Buran is divided into a before and an after, in between a single 3 h 25′ 22 s flight on November 15, 1988. On that day, after circling the earth twice, the Buran spacecraft landed at the Baikonur Cosmodrome. Despite the high expectations for it, Buran never flew again. And it went down in history as the Soviet Union's most expensive and important space exploration project, cancelled soon after the dissolution of the USSR. But it was nonetheless the moment of highest technological innovation in the history of the Soviet Union. Over the years, more than 1 million people, 1286 industries and 86 development departments had worked to complete the project.

The orbiting module had only two engines needed for manoeuvres in orbit or during re-entry. It was also larger than the Shuttle, with a wingspan of 24 m, a length of 35.4 m and a height of 16.5 m. It was so massive and heavy that it had to be carried to the launch pad horizontally, then placed vertically for launch.

The most remarkable innovation was its fully autonomous navigation system, a miracle of technology at the time. This was all because of the Soviet tendency not to trust individuals (Gagarin was not told how to disarm the autopilot), or because of high-risk military scenarios. From a Western and, even more so, a US perspective, the level of automation of Soviet space equipment at the time may seem incredible, as well as the massive use of computer controls, far superior to those used by nations (such as the US) who could boast among their possible suppliers of computer technology famous names like IBM, Digital Equipment Corporation, Honywell, Univac, and Hewlett Packard, not to mention Cray, with its vector supercomputers.

Along with the evolution of the Buran, the designers were also studying the carrier rocket that would carry it into orbit. Thus was born the design of the Energiya rocket, the most dissimilar aspect of the Shuttle. Some military analysts explained to *New Scientist* magazine that the Buran had no civilian mission in mind and was only designed to carry weapons into space!

Returning to the rocket, the Soviets instructed Valentin Glushko's Design Bureau to come up with a response plan. They came up with a new OKB-1 launcher, later to be called the Energiya.

On that day in November 1988, the Energiya carried the Buran on its back. The power of the rocket made the earth tremble for miles as the spacecraft disappeared into the absolute darkness of the night on the steppe. The Buran made two unmanned orbits, re-landing fully automatically after 206 min of flight and having lost only 8 of its 38,000 thermal tiles (much better than the Shuttle, which was always plagued by this problem). After the demonstration of its capabilities, the Buran did its swan song. The Soviet Union would crumble within a couple of years, and, in 1991, Boris Yeltsin would put an end to this most expensive and ambitious Soviet space programme. But Buran was, above all, a totally useless programme. The spacecraft was conceived solely as a military response to the Shuttle, and was never part of a wider space or satellite research programme (think of what the Shuttle did for the Hubble Telescope[2]). However, destiny had not finished inflicting itself on the Buran: the flight shuttle was kept in a museum in

---

[2] The Hubble Space Telescope was brought into orbit by the Space Shuttle Discovery in 1990, after which no less than five Shuttle missions have repaired, upgraded, and replaced various systems on the telescope, including all five of its main instruments.

Baikonur, together with the Energiya rocket, but one fine day in 2002, the ceiling of the hangar decided to put an end to the Buran for good, when it collapsed and demolished everything. Yet, there in the steppe, in a hangar some 30 kms away from the old launch pad, there is still a Buran, number 1.02, 95% complete at the time of the programme's suspension. It was redis-covered by photographer Ralph Mirebs in June 2015. Inside the hangar, more than 60 m high, there is also another, less complete Buran; indeed, others were under construction, and the one considered to be in the most advanced stage (30–50%) is still stored at Zhukovsky International Airport. The photos by Mirebs are an absolute must-see, and perhaps most shocking is the fact that, despite never having flown, the shuttles are broken and dented, their internal mechanisms having largely been looted by thieves. Others have followed in the footsteps of Ralph Mirebs and even made videos.[3] It seems that, even now, there are still those who organise clandestine tours to visit the hangar and see the Burans live: it takes a good deal of "courage," orienteering skills and enough training to walk (at night) a few dozen kilometres into the Kazakh desert that is home to the huge Baikonur cosmodrome.

Buran and the Shuttle are just two parts of two complex launch systems. The fundamental difference is that, while the Shuttle aimed to reuse its main engines by placing them on the spacecraft, the Buran completely delegated the job of climbing into orbit to the Energiya rocket. And thanks to the Energiya, Buran was able to beat the Shuttle in the payload category: it could carry about 30 tonnes into low orbit (as opposed to the Shuttle's 25 tonnes) and return 25 tonnes to earth, while the Shuttle could return only 14.4 tons to earth. In addition to this, the Buran could take off like an aeroplane, even though both shuttles could be moved from one point to another in the terri-tory, the US Shuttle on top of a Boeing 747 (the well-known Jumbo) and the Buran on top of an Antonov An-225 Mriya, the largest aircraft in the world. Those who visited the Paris-Le Bourget International Aeronautics and Space Show in 1989 were able to witness the demonstration flight of an An-225 with the Buran space shuttle attached to its back.

While the shuttle's boosters ended up parachuting into the warm seas off Florida, the Energiya's boosters, designed to launch from the Kazakh steppes, also had retrorockets to brake them at the last moment. The central stage was destined for the Pacific, but there were plans to evolve it so as to make it reusable.

The basic element of this automation, however, was greater safety in the event of launch anomalies; the Soviets thought that, in the event of

---

[3] https://www.youtube.com/watch?v=-q7ZVXOU3kM.

a faulty re-entry from orbit, it might be useful to have a minimum of "room to manoeuvre" for corrections. So, 500 potential scenarios were pre-programmed into the memory of the computer and the Buran would be limited to identifying the right one, executing it on time and bringing back the spacecraft to Earth safely.

It was a pity that, at the end of the 1980s, the Buran project stopped due to a lack of money, even though it had achieved all its objectives. Such a vast programme had cost as much as 14.5 billion roubles, which today would be $19 billion! And the Soviet Union was not in good shape at that time—on the contrary! However, even though the Energiya has only flown twice, its rockets, in their various versions, still fly on the Atlas or Zenit.

Which one is better? The Buran or the Shuttle? The Shuttle has 135 missions to its credit, it saved Hubble, it built the ISS, it is a vehicle that has made space history. Of course, it has never achieved many of the (cost) objectives that had been set for it (for example, the forecast of one flight a week has turned into one flight every three months or so), and it has had two statistically significant fatal losses. But the shuttle is a reality. The Buran flew only once and only proved that it could do so; moreover, it needed no crew. So, it remains the most expensive exercise in style in human history. Perhaps the Buran could have surpassed the Shuttle in safety: automation, a heat shield that won't come off, liquid fuel engines, etc., all things to consider. But the cost and orbital capabilities have never been tested. It would have been nice to see them measured up there. Perhaps it would have been even more amazing to see them cooperate. Utopia![4]

5.5   Russians the First to Land on Mars

Will it be so? If we are referring to missions with humans on board, we don't know yet! Many countries are planning missions.

Attempts to explore Mars began in the 1960s using unmanned automatic probes; about two-thirds of the missions have failed, and the orbiters, landers and rovers designed to arrive on the planet, touch down on Martian soil, and explore it, including collecting soil samples, taking pictures and sending them back to Earth along with other environmental, meteorological and physico-chemical data, have all been lost or destroyed.

The first to try were, again, the Soviets, who, in October 1960, launched two probes within days of each other with the aim of flying over the planet.

---

[4] https://rollingsteel.it/aerei/buran-space-shuttle-unione-sovietica-urss/ *di Paolo Broccolino 18-6-2021.*

Both missions failed and did not even reach Earth's orbit. The USSR tried again in 1962; the Mars 1 probe left Earth's orbit and began sending data on the interplanetary environment. On March 21, 1963, communications stopped when Mars 1 was 106,760,000 km from Earth. However, it was the US that succeeded, with Mariner 4, which was launched on November 28, 1964, and orbited Mars on July 14, 1965, sending back 21 photos. To demonstrate that the US-USSR competition was also extreme in reaching Mars, one need only consider that, on November 30 1964, only two days after Mariner 4, Zond 2, the fifth Soviet probe to attempt a flyby of Mars, was launched.

No photographs or illustrations of the probe have ever been published.

In May 1971, there was a repeat of the close launch of Mariner 9 (USA) and Mars 3 (USSR). The US flew to Mars at almost the same time as the Soviets. Using technology from half a century earlier, the USSR managed to land the Mars 3 mission lander on the surface of Mars in a fully automated manner. The landing module separated from the orbiter just over 6 months after launch, on December 2, 1971, 4 h and 35 min before the probe reached Mars, and entered the atmosphere of the red planet at 5.7 km/s. After a perfect flight, the module touched down on the Martian soil and became immediately operational. While a dust storm around the planet hampered the collection of US data, the Soviet Mars 3 probe not only managed to land, but also started sending back data. Too bad it was only for 14 s! Then, silence.

The technicians in the control centre were amazed… for a few moments!

Mars 3 ceased communication, disappointing Moscow's expectations of redemption after 1969. Before it stopped, however, Mars 3 had one sound victory! It sent back the first photo of the surface of the red planet, at close range, although blurred, as a storm was raging.

It must be said that, in May 1971, the Soviets had launched not one, but two probes for the Red Planet: Mars 2 and Mars 3, each equipped with a landing module. The first reached Mars in late November 1971, the second on that historic December day. Both ran into a storm of debris surrounding the Red Planet and, as mentioned above, only Mars 3 made it there, albeit briefly. No firm evidence was ever found that the failure (or Mars 3's short life) was due to the storm, and the USSR did not want to talk about the result of those transmissions from Mars, which appeared to be images of dark lands and skies. Mystery. It was only on December 31, 2012, that the Mars Reconnaissance Orbiter, which was following the movements of two NASA rovers on Mars from above, flying over Ptolemaeus crater, the very landing site of Mars 3, managed to take photos of mechanical remains attributed, with certainty, to the Soviet probe.[5]

To this day, only speculation can be made as to why the 14-s connection was made. Two telephotometers were installed on the lander, working in different bands of the spectrum (telephotometers were photographic instruments that sent panoramic images of the surface, scanning them line by line like a fax machine). It is strange that both stopped working at the same time and after only a few seconds. One possible explanation was an electrical phenomenon called the **corona effect** (whereby an electric current flows between a conductor with a high electrical potential and a surrounding neutral fluid, usually air). It had already happened in World War II, when British radio operators experienced malfunctions in their transmitters due to a corona discharge while working in the Lebanese desert during a sandstorm. On Mars, the dust particle size, humidity and atmospheric pressure are much smaller, but the wind speed is much higher than in the Lebanon desert. Perhaps the corona effect was the reason why the signal from Mars suddenly disappeared. Even if the radio connection worked well, it would have been impossible to receive topographic images in the middle of the sandstorm. In practice, the Martian air, saturated with suspended dust from the global storm and made particularly turbulent by the wind, had created electrical interference. This prevented the radio signals sent by the lander from reaching the orbiter in space, and, from there, Earth. Unfortunately, the storm continued

---

[5] SOME TECHNICAL DATA: In the picture, you can identify on the ground: parachute, retractor, lander and heat shield. Interesting details can be deduced: the parachute had a diameter of 7.5 m, the retractor that landed Mars 3 shows a linear extension that resembles the chain that held them attached. The landing capsule consisted of four triangular "petals" that, when opened, would straighten the lander and release the rover PrOP-M, with cameras with 360-degree views, a mass spectrometer to study atmospheric composition, temperature, pressure and wind, and devices to measure the mechanical and chemical properties of the surface and to search for organic compounds or signs of life. PrOP-M weighed 4.5 kg and was remotely piloted via a cable as long as 15 m. The rover was designed to be "disembarked" from a mechanical arm of the lander and to move into the cameras' field of view, taking measurements every 1.5 m. Two small metal rods acted as "sensors" and allowed the rover to scan small obstacles in its path automatically, since any commands from earth would take too long to reach the device. Four antennas, which protruded from the top, would ensure communication with the orbiter. Due to the loss of the lander signal, it was never known if the PrOP-M had actually landed and performed any operations. The landing module had been sterilized prior to launch to avoid any contamination of the Martian environment. The heat-supply shield is the only object that the debris storm almost completely submerged. The U.S. sent those data to Arnold Selivanov, one of the Mars 3 creators, and Vladimir Molodtsov, a former Russian space engineer, to integrate them with those kept in the archives of the Moscow space agency. It seems that there was a loss of fuel during the mission: this prevented Mars 3 from placing itself in the planned orbit of Mars, lasting 25 h, allowing it to arrive only in an orbit lasting 12 days and 19 h. Mars 3 sent to Earth data that showed mountains up to 22 km high, hydrogen and oxygen in the upper atmosphere, and very high temperature ranges between −110° and +13° Celsius. The dust raised by the Martian storms could reach a height of 7 km.

for months on end, putting a premature end to the scientific and topographical surveys that the automated Martian station was to carry out when it landed.

However, this does not detract from the great achievement of the Soviet team of scientists and technicians, who, despite enormous political pressure and limited time, succeeded in brilliantly solving the problem of how to land a lander on the surface of Mars in a fully automated procedure. The soft landing of the Mars 3 lander in December 1971 was remarkable, if one considers that the feat was carried out under prohibitive conditions, without precise ephemerides on the position of Mars, without a thorough knowledge of the conditions of the Martian atmosphere and soil, and with a continuing global dust storm, which made the surface of the planet completely invisible.

In particular, it may seem unbelievable, given that the transmission from the ground by Mars 3 lasted only 14 s, but the data sent by Mars 2 and Mars 3 over the entire mission made it possible to measure the magnetic field of Mars; determine the temperature and atmospheric pressure at the planet's surface; identify the nature of the surface rocks visible along the orbits of the two probes; define the density, thermal conductivity, dielectric permeability and reflectivity of the soil; determine the temperature of the lower atmosphere and its variations according to time of day and latitude; establish that the concentration of water vapour in Mars' atmosphere is 5000 times lower than that in the Earth's atmosphere; define the extent, composition and temperature of Mars' upper atmosphere; and determine the height of the dust clouds (around 10 km) and the typical size (a few microns) of the particles of which they were composed.

Exploration continues, and the next "launch windows" to reach Mars are scheduled for October–November 2022 to April–May 2023 and December 2024-January 2025 to July–September 2025. Launches with humans on board cannot be envisaged in any of these cases.

5.6   Today on the ISS

The Space Station is a shining example of how international differences fade into the background when you have a big goal, a common passion. (Samantha Cristoforetti)

5.6.1 The International Space Station

The name already explains the splendid project. As a brief chronology, we note that construction began in 1998, and 2024 should be the final date of its use, with 2028 being the year by which it should be dismantled, destroyed or partially reused. But what to do with it afterwards? In 2020, Vladimir Solovjev, deputy director of the planning company 'RKK Energija', said:

> There are already a number of severely damaged elements that are going out of service. Many of them are not replaceable. After 2025, there will be an avalanche failure of numerous elements.

The ISS was a cyclopean project, costing a staggering €100 billion, spread over 30 years.

What are its functions, one wonders? Here is what it does:

- It develops and tests technologies for space exploration,
- It keeps a crew alive on missions beyond Earth's orbit and facilitates experience for long-duration spaceflight,
- It serves as a research laboratory in microgravity environments, where crews conduct experiments in biology, chemistry, medicine, physiology and physics and make astronomical and meteorological observations.

It represents a huge set of tasks, established by agreements between the various governments. The station is served by the *Soyuz, Progress, Dragon* and *Cygnus* spacecraft and the *H-II Transfer Vehicle*. International cooperation!

By 2020, it had been visited by more than two hundred and forty astronauts and cosmonauts from eighteen different countries.

5.6.2 The Station's Origins

The International Space Station is the amalgamation of several national space station projects that originated during the Cold War.

In the early 1980s, NASA planned to build the *Freedom* station as a counterpart to the Soviet space stations *Saljut* and *Mir*. However, *Freedom* did not make it past the design stage; and after the fall of the Soviet Union and the end of the Cold War, its implementation was cancelled.

In addition, NASA and all other space agencies were in dire financial straits. On the one hand, this was a good thing, because it convinced the US administration to contact other governments for a joint project. In the meantime, the economic chaos in post-Soviet Russia also led to the cancellation of the construction of the *Mir-2* space station (which was to have succeeded *Mir*).

This led to a series of advantageous agreements.

Here are the next steps taken by the various governments:

- In the early 1990s, the US government began to involve space agencies in Europe, Russia, Canada and Japan in the project of building a joint space station, referred to as "*Alpha*."
- In June 1992, US President George Bush and Russian President *Boris Yeltsin* made official agreements to collaborate on space exploration.
- In September 1993, US Vice President Al Gore and Russian Prime Minister Viktor Černomyrdin announced plans to build the space station. The *Shuttle-Mir* Programme was also launched, leading to joint missions of the US space shuttle to the Soviet space station *Mir*. In this way, collaboration between the Russian and US space agencies was increased and solutions were tested to integrate Russian and US technology into the ISS.

In the agreements between the international partners, it was foreseen that the solutions that each agency had developed for its own space station would be re-used; thus, the station ended up being based on NASA's *Freedom* station designs, the *Mir-2* station (which became the heart of the *Zvezda* module), ESA's *Columbus* laboratory and Japan's *Kibo* laboratory.

5.6.3 Purpose

Our daily lives are filled with the results of space activities. The weightlessness of the space environment allows for experiments that are impossible on Earth and, moreover, of long duration. Since the next missions to the Moon and Mars will presumably use orbiting stations, the experience gained on the ISS seems crucial.

Notwithstanding international cooperation, both among the nations that contributed to the creation of the Station and those participating in it, part of the aims, and therefore of the activities of the astronauts on board, also concerns education and teaching: in fact, the ISS crew allows students on Earth to carry out demonstration experiments.

5.6.4 Scientific Research

Research fields explored on the ISS include: human studies, space medicine, biology (including biomedical and biotechnology experiments), physics (including fluid mechanics and quantum mechanics), materials science, astronomy (including cosmology) and meteorology. With the 2005 *Authorization Act*, NASA designated the US segment of the ISS as a national laboratory, with the aim of increasing its use by other federal agencies and the private sector.

Research on the ISS has improved various areas of knowledge: in particular, the effects (still little known) of a long stay in space on the human body. Studies have focused on muscle atrophy, fluid dynamics and bone loss. The data will be used to determine whether both space colonisation and long-duration human flights are feasible.

As of 2006, the data on bone and muscle loss are not encouraging. There would be a significant risk of fractures and circulation problems if astronauts landed on a planet after a long interplanetary journey (e.g., a journey lasting at least six months, which is as long as it takes to reach Mars).

There is usually no doctor on board the ISS. So, diagnosing health conditions is a challenge. But remotely guided ultrasound will have applications on Earth both in emergency situations and in the open countryside, where it can be difficult to get the care of an experienced doctor.

Researchers are also studying the effect of a near 0-G environment on the evolution, development, growth and internal processes of plants and animals. NASA aims to investigate the effects of microgravity on the synthesis and growth of human tissues and unknown proteins that can be produced in space.

Studies in microgravity on the physics of fluids will allow researchers to better understand a huge range of factors. Since fluids in space can be mixed almost completely regardless of their weight, space is the ideal place to study combinations of liquids that would not mix at all on Earth.

Then, there are the experiments carried out outside the station, at very low temperatures and in conditions of near weightlessness: thanks to them, we will expand our knowledge of the states of matter (superconductors above all). The combination of the two factors—very low temperatures and near weightlessness—should show the state transitions as if we were seeing them in slow motion.

On the ISS, another important research activity is the study of materials. In space, even foam and flames grow and behave differently. Other areas of interest include cosmic rays, cosmic dust, antimatter and dark matter in the universe. Among the experiments being conducted is the measurement of solar energy production in recent years. These data are important because changes in solar activity can be reflected in the Earth's climate. Various materials are also exposed to the harsh space conditions outside the Station.

In addition, there is great anticipation for the *Alpha Magnetic Spectrometer*, a detector installed on the ISS used for the study of particle physics. It is designed to find new types of elementary particles through high-precision measurement of the composition of cosmic rays.

Important: the ISS **does not, we repeat does not, hunt for UFOs**. Elisa Nichelli, the astrophysicist and scientific journalist, wrote about the "incident" in an article posted on the INAF website on July 15, 2016.[6]

Once again, someone pointed the finger at NASA: they allegedly turned off the live feed just after a UFO appeared on the horizon.

How did the plotters' accusations against NASA begin? In the beginning, it was a YouTube video that did the trick. The user StreetCap1 rightly recalled the meaning of the term UFO, "Unidentified Flying Object", but then provocatively added that the broadcast had stopped just when the UFO stopped. And this is not the first time that there have been insinuations about the videos of the live broadcasts (High Definition Earth Viewing System), made using four high-definition cameras, placed at strategic points on the ISS. These transmissions stop in two cases:

- When there is no signal
- When the ISS is in darkness, because it is passing through the Earth's cone of shadow.

NASA spokesman Daniel Huot explained:

The station regularly passes out of range of the Tracking and Data Relay Satellites, the system used to send and receive the video signal. If the signal is lost, the cameras show a blue screen (indicating no signal) or a preset image.

You only have to watch the live feed for a few hours to realise that the lack of a signal happens often: every half hour, the ISS makes a complete circle around the Earth. Daniel Huot went on to say:

As far as the 'object' is concerned, it is very common for things such as the Moon, space debris, reflections from the station windows, parts of the ISS structure itself or lights from Earth to appear in the images and create artefacts in the photographs and videos collected by the station.

Paolo Attivissimo, journalist and hoax hunter, explains to Media INAF:

Those who believe that NASA sees UFOs all the time do not know how the technologies work! In addition, NASA has little money at its disposal (the Houston Control Centre looks like a university campus that has not improved since the 1960s). They are people who could not keep a secret, also because there is a pervasive culture of transparency at NASA (a decades-old habit, created to counter the obsessive secrecy of the Soviets). If you are a friend of theirs, they will tell you gossip and tidbits with relish: imagine keeping secrets! Besides, NASA has only to gain from announcing the discovery of extraterrestrial life: there would be funding pouring in.

He concludes:

No one at NASA denies the possibility that extraterrestrial life exists, but it must be demonstrated that such life is coming to visit us. And you have to ask yourself what logic it would make for a cosmic traveller to play hide-and-seek. What the ufologists are describing is the equivalent of a tourist taking a long plane ride, arriving at his destination, popping into the airport to photobomb a cashier, and then bailing. What's the point?

5.6.5   Orientation and Altitude Control

The ISS is maintained in a nearly circular orbit, with a minimum altitude of 278 km and a maximum of 460 km. It travels at an average speed of over twenty-seven thousand kilometres per hour and completes 15.7 orbits per day.

The station is constantly losing altitude due to minor atmospheric friction, so it needs to be returned to higher altitude several times a year. It takes about two orbits (three hours) to complete the altitude increase. The station's orientation is maintained through electrically-powered control gyroscopes.

Continuous on-board activities, crew changes, the arrival of various spacecrafts for loading/unloading supplies and waste, necessary maintenance, etc., have caused numerous alarms over the years. There have been air leaks, ammonia leaks, electrical short circuits, smoke on board, collisions with unidentified objects and other mechanical or electrical anomalies, so far always minor, sometimes caused by faulty sensors. However, in many cases, moments of panic were experienced before the cause was identified. Bacteria and fungi have also been detected on board, some of which can even corrode metal!

Here is one of the headlines that recently appeared in the international media: ISS alarm, Russian Nauka module out of control! Emergency manoeuvres with 7 astronauts ready to leave the station.

In this case, the docking of the Nauka module on the ISS had caused a loss of orientation of the entire station due to a sudden ignition of the thrusters. The station tilted alarmingly.

But what is Nauka (Russian for 'Science')? It is a spacecraft that, once in orbit, can fly itself to the station to attach itself, which no US or EU module can do. It was completed after 14 years of work and many delays caused by technical problems. It was finally added to the ISS in July 2021 and is Russia's first scientific laboratory in space. Among other things, Nauka will be the largest and most complex module in the entire Russian segment of the International Space Station. Nauka, where experiments on growing embryos will be carried out, is very large. Inside, there are 21 workstations. And there is the laboratory for experiments and a centrifuge to recreate the force of gravity. Finally, an automated 11-m-long "space arm," which is intended to avoid or reduce "space walks" for repairs on the Russian side of the ISS, will be able to move loads of up to 8,000 kilos with an accuracy of 5 mm.

Unfortunately, Nauka has been one of the most problematic projects, a never-ending set of problems. Russia started it in the early 2000s, not even from scratch, but rather by transforming the backup version of Zarja into a scientific module. When it was almost ready, it was decided that it would be turned into a space laboratory. The launch was planned for 2007, but was constantly delayed due to technical and financial problems. The worst difficulty came in 2013: during inspections, metal shavings as small as 100 microns were found in the fuel lines and tanks. Nauka's technicians and engineers explained:

> Rinsing was a terrible hassle. We have been working seven days a week in two shifts, with constant rinsing and testing. We manage to get a certificate that the tank is clean, and after a while, there is new contamination.

The shavings could have killed Nauka: even if they were tiny, foreign bodies in the tanks could get into the engines and shut them down. Thus, the module could stay in orbit and catch fire in the atmosphere. The tanks could not even be changed, as the manufacturer was from the Soviet period, since closed and scrapped. Then, the infernal shavings were also found to have passed into the reserve tanks. After the manoeuvres to clean everything up, the commission finally gave the go-ahead: but only to use it once, and only as long as Nauka was never part of the ISS's power supply and movement, so as not to jeopardise its safety. In July 2021, when Nauka had been added to the ISS only a few hours earlier, there was the trouble referred to above in some media headlines, which "derailed" the station itself. Nauka had suddenly started up and had rotated the ISS by 45 degrees! Evidently, the curse of the Nauka continued. As many as 7 astronauts had to perform rearrangement manoeuvres, which had never needed to be done before. There was quite a bit of suspense in Houston and Moscow as the station tilted and rotated itself for almost an hour. The Russian space agency gave this explanation: it was a "short-lived software failure," which sent the order to fire the thrusters. So, the rotation began, and ended when Nauka's fuel ran out and the ISS was refitted. Meanwhile, on the ground, everyone was ready to prepare SpaceX's CrewDragon Endeavour, in case of evacuation. Live coverage of the ISS was interrupted, and there were communication blackouts. From the control room, US astronaut Drew Morgan asked his colleagues in orbit to look out of the dome and other portholes to see if there was any debris around the ISS. There was.

Before the live feed stopped, the station was surrounded by a cloud of small debris, caused by the burning of fuel from the engines of both Nauka and Zvezda, another Russian module, which was turned on to balance the thrust of the new laboratory. The orders were: cosmonauts Novickky and Dubrov were to use the engines of the Soyuz Progress MS-17 to bring the ISS back into position.

Bernardo Patti, who heads the Esa Exploration Programme, said:

> At no time was there any real danger to the life of the crew on board the Space Station. The module had registered several anomalies from the start and the Russian space agency will now carry out its own investigation to ascertain the details of the causes.

But three hours later, Nauka's engines suddenly started up again, as

> ...some propellant control valves had not closed.

What was the reason for the failure? The most credible hypothesis is that Nauka ignited due to a failure to update its software: ergo, the module thought that it was still in flight and not already docked to the ISS.

The fact is that, at first, nobody noticed anything. Only the station's computers had noticed the rotation and, as a reaction, had given a reverse impulse via the Zvezda module. That was not enough. The ISS continued to rotate up to 45 degrees, when it was decided that the engines of the Russian Progress Module should also be switched on.

5.6.6 Computers

The International Space Station is equipped with about one hundred portable computers, made to work in weightlessness.

But there was a "mystery" concerning the (ground) computers.

On Friday, March 2, 2012, some media reported that a PC with secret ISS codes had disappeared.

This was a shocking theft at NASA. The American space agency had completely lost track of a PC containing codes for the command and control of the International Space Station. An alarming affair, dating back to 2011, made public on February 29 of the following year, during a committee meeting at the US House of Representatives. NASA Inspector General Paul K. Martin spoke about it in front of the Science, Space and Technology Committee, explaining that the PC had been stolen twelve months earlier. This led to the loss of algorithms used to control the space station itself. But this was not the first theft! It was the 48th such theft between 2009 and 2011, which is of great concern to NASA. Inspector Martin also pointed out that the number may be underestimated. The stolen PCs contained confidential data on intellectual property and personal data, as well as social security numbers and data on the Constellation programme and the Orion spacecraft.

So far, there have been, among other things, thousands of attempted interferences by foreign intelligence services or hackers.

But through the Office of Public Affairs, NASA released the news that:

> International Space Station operations are not put at risk by the data breach. NASA has made significant progress in improving the protection of computer systems.

According to Martin, the US space agency, which is very sensitive territory and prone to cyber attacks, suffered 47 hacker assaults in 2011 alone. Thirteen of these were successful in bringing down computers. All of these offensives are a part of the 5,408 hostile incidents and intrusions that attacked NASA's IT security systems between 2010 and 2011. Malware was also installed. In total, the failures cost $7 million. Inspector Martin pointed out that

The intrusions could be sponsored by foreign intelligence services.

5.6.7   Life on Board—Crew Activities

At night, the windows are covered to give the impression of darkness, as the sun rises and sets sixteen times a day on the station. A typical day for the crew starts with the alarm going off at 06:00, followed by a general inspection of the station. The crew has breakfast and takes part in a daily planning briefing with Mission Control. Work begins at 08:10. The lunch break begins at 13:05 and lasts one hour. The afternoon is dedicated to various activities that end at 19:30, with dinner and a briefing. The astronauts go to bed at 21:30. In general, the crew works ten hours a day on a weekday and five hours on Saturdays, with the rest of the time devoted to rest or unfinished work.

After the Columbia accident (February 1, 2003) and the subsequent suspension of the Space Shuttle programme, the future of the ISS remained uncertain until 2006. In fact, in July 2005, immediately after the launch of Shuttle Discovery on the STS-114 mission, several problems arose that were solved with impromptu repairs in open space.

NASA decided to suspend the space programme again until the problems were resolved. During the interruption of the shuttle flights, the station survived only thanks to supplies from the Russian Soyuz spacecrafts. The crew was reduced to two from the three on the flight plan. The Shuttle did not visit the station for a long period: this posed many problems, since construction was halted (the Shuttle was the only spacecraft able to carry the main modules into orbit) and operations themselves were limited by the presence of waste not brought back to Earth. The Progress transports and the STS-114 mission made it possible to reduce the waste problem.

There is an interesting piece of news about the first (and, so far, only) protest in orbit. It is the story of the mutiny of the crew of the Skylab4 mission. The reason? NASA had imposed too much work![7]

"Mutiny" may be an exaggeration, but it was certainly a rebellion. Three astronauts in orbit were fed up with the exaggerated demands coming from the ground. They "went on strike" by refusing to do anything else. You would no doubt be interested to know the names of those who did it. Here they are:

---

[7] https://www.focus.it/scienza/spazio/sciopero-in-orbita.

Gerald Carr, William Pogue and Edward Gibson, three astronauts from the Skylab 4 mission, launched in 1972. There was a lot of gossip and few official statements, but it certainly forced NASA to review the various tasks, rethink long-duration flights, taking into account the psychological aspect, i.e., stress, and possible disagreements.

Skylab4 was the third mission to take a crew to the American space station, the laboratory put into orbit with the aim of studying the feasibility of life in space, as well as making a wide variety of scientific observations.

The previous mission went smoothly, with the crew staying on Skylab for 59 days and doing even more than the required work. Skylab4, which left on November 16, 1973, also had a very demanding schedule. But things got off on the wrong foot right from the start.

As soon as they were in orbit, Pogue began to exhibit signs of space sickness. The astronauts thought that it was just a temporary thing (as indeed it was: Pogue would not die until 2014, at the age of 84), and thus did not inform Earth. But the Control Centre had heard the discussions, so the doctors scolded the astronauts for keeping quiet.

5.6.8 Hooked to the Treadmill—Exercise

The worst effects of prolonged weightlessness are muscle atrophy and space-flight osteopenia. Other important effects include fluid redistribution, a slowing of the cardiovascular system, reduced production of red blood cells, disturbances in balance and a weakened immune system. Minor symptoms include loss of body mass, nasal congestion, sleep disturbances, excessive flatulence and facial swelling. These effects quickly disappear upon returning to the earth.

To avoid some of these problems, the station is equipped with two treadmills, some weight-lifting equipment, and an exercise bike. Each astronaut spends at least two hours a day exercising, using bungee cords to hook up to the treadmill.

For the researchers, exercise is good protection for bones and limits the loss of muscle mass in those who spend long periods of time in the absence of gravity.

5.6.9 Hygiene

There are no showers, and crew members must wash themselves with a jet of water, wet wipes and soap dispensed from a tube. Astronauts also have shampoo and edible toothpaste to save water. There are two toilets on the ISS, both of Russian design. Solid waste is collected in individual bags, stored in an aluminium container. Once the containers are full, they are transferred to the Progress spacecraft for disposal. Liquid waste is collected and transferred to the water recovery system, where it is recycled as drinking water.

5.6.10 Food and Drink

As you have ascertained by this point, were are not talking about a five-star hotel. The food eaten by the station crews is frozen, refrigerated or canned. Before the mission, the menus are studied by the astronauts, with the help of a dietician. Since the sense of taste is reduced in orbit, spicy food is favoured. Each crew member has individual food packages and cooks them in the on-board kitchen, which is equipped with two food warmers, a refrigerator and a dispenser of both hot and cold water. Drinks are provided in the form of dehydrated powder, which is mixed with water before consumption. Drinks and soups are sipped from plastic bags with straws. Solid food is eaten with a knife and a fork, attached to a magnetic tray. Food crumbs and fragments must be collected to avoid clogging the station's air filters.

5.6.11 Sleeping in Space

The station is equipped with sleeping quarters for each permanent crew member, with two "sleep stations" in the Russian segment and four more in the US *Tranquillity* module. The US quarters are soundproofed single cabins, where a person can sleep in a sleeping bag, listen to music, use a laptop computer and store personal items in a large drawer or in nets fixed to the walls. The accommodation also provides a reading lamp and a shelf.

Visiting crews sleep in a sleeping bag attached to the wall. The crew quarters are well ventilated, otherwise astronauts could suffocate due to the carbon dioxide bubble that can form around them.

5.6.12 Radiation Exposure

Without the protection of the Earth's atmosphere, astronauts are exposed to very high levels of radiation from the constant stream of cosmic rays. In fact, station crews are exposed to about 1 millisievert of radiation per day. This is the same amount that every human being on Earth is exposed to, but, on Earth, they receive it over the period of a year. Astronauts therefore have a much higher risk of chromosome damage to lymphocytes. These cells are essential for the immune system, and damage to them may contribute to the low immunity experienced by astronauts. Increased exposure to radiation leads to a higher incidence of cataracts. Protective drugs and shielding can reduce the risks to an acceptable level, but the amount of data is scarce, and long-term exposure will increase these risks.

Compared to previous stations such as *Mir*, radiation shielding on the ISS has improved, but further technological advancement is needed to enable long-duration human spaceflight in the solar system. Radiation levels on board the ISS are now (only) five times higher than for airline passengers.

In their free time, astronauts can observe our planet from space. The beauty of this view is incredible. All those who have had the opportunity to experience it have said unanimously: if, at least once in their lives, people could observe the Earth from up there, they would certainly take better care of their home.

The so-called "dome," the module of the International Space Station designed in Europe, was built in Italy by Thales Alenia Space and was added to the station in 2010. Those who have visited it say that it is certainly the most beautiful "room with a view" in the (known) Universe.

5.6.13 Sabotage! the Mysterious Hole on the ISS[8]

In 2018, there occurred the famous "thriller" about the hole that someone had drilled on a Soyuz spacecraft docked to the International Space Station.

How did it happen? The astronauts on board realised that the internal pressure was dropping, slowly but steadily. So, they began to inspect every inch of every chamber on the Station, eventually discovering a tiny 2 mm diameter hole in the wall of the Soyuz Ma-09 spacecraft. A tiny hole, sure. But if no one had spotted it, the astronauts' safety would have been at risk. In practical terms, it was easy to fix. Epoxy resin and tape, and voila! All fixed.

---

[8] https://www.focus.it/cultura/mistero/sabotaggio-iss-2018-colpevole-astronauta-nasa.

https://www.wired.it/scienza/spazio/2019/09/23/mistero-buco-iss/ by Mara Magistroni.

But it could not be hushed up. Who the hell had made the hole? First of all, the astronauts did a spacewalk to monitor the outside of the ISS and realised that the hole was not caused by space debris that had crashed into the station. What's more, the pattern of the hole indicated that it was human-made. The photographs showed a hole that appeared to have been made by a drill. When the hole was closed, one fact became apparent: it had been drilled into that part of the Soyuz shuttle that was detached before the crew module began its return to Earth. If it was human-made, it was carefully planned so that it would only affect the ISS, but not the crew on the return journey. It was determined that the hole had been drilled from inside the spacecraft with an uncertain and trembling hand. This meant that it had been done in zero gravity, so not on Earth, not before the Soyuz had left.

After more than a year, no one knew who was to blame for the sabotage. Or perhaps it was a mistake? But Dmitry Rogozin, head of the Russian space agency Roscosmos, said, in 2019:

We know very well what happened, but it will remain a secret.

So, Nasa demanded an answer.

But what happened was something that no one expected. An article in TASS reported some US complaints about the work of Roscosmos, after which an official assigned to investigate, who did not want to be named, identified the culprit as American astronaut Serena Maria Auñón-Chancellor, the only woman on board the ISS when the hole was discovered. The reason? The anonymous official reported that Auñón-Chancellor had been complaining of thrombosis in her jugular vein since she arrived in space, which she dealt with when she returned home. The abnormal health condition would have prostrated the astronaut, so she allegedly found a way to cause a malfunction small enough not to endanger anyone's life, but serious enough to evacuate the ship. By coincidence, the official reported, the video camera that was filming the passage between the American module and the Russian spacecraft had broken down at the very time that the hole incident occurred.

NASA's head of human spaceflight, Kathy Lueders, did not wait to respond. At a press conference, she told reporters that it was an absurd accusation against NASA astronaut and flight engineer Auñón-Chancellor. In short, pure delirium.

Serena is an extremely respected crew member who has served her country and made invaluable contributions to NASA.

The complaint and response have deepened the chill between NASA and Roscosmos. But no one has yet provided a plausible explanation for what happened. Other rumours speak of human error during the construction of the spacecraft. Russia has yet to confirm this. So, why does Rogozin say that he knows, but won't talk? In the Houston Chronicle, NASA Administrator Jim Bridenstine said that he had spoken with Rogozin and reiterated that Russia has officially given no explanation whatsoever. It is evident that NASA does not want to come to blows with Roscosmos, but the episode must be clarified.

Appendix: Short Excerpts from Kamanin's Diaries

Abstract These diaries are available only in part and in English. A few paragraphs that specifically affected the episodes in this book have been taken from them. Kamanin often speaks of himself in the third person and with scrupulous care for historical truth. He rarely indulges in personal commentary. These, then, are a series of faithful stories of what was happening at Star City between August 1962 and December 1963.

Diary (s.m.). Daily documentation of a part of one's life that one can tell oneself without blushing. (Ambrose Bierce)

Translation by Mark Wade[1]

1962 August 15—Landing of Vostok 4.

Recovered August 15, 1962 6:59 GMT. Landed 48:09 N 71:51 E. By 07:00 the temperature aboard Vostok 4 is down to 10 °C, and the humidity at 35%. Popovich is ready to continue for a fourth day, but he admits the cold is getting to him. Keldysh and Rudenko now support returning Vostok 4 to earth on the 49th orbit, but Smirnov still wants to go for the extra day. Then Popovich radios 'I observe thunderstorms (groza). Groza is the pre-agreed code word to indicate that the cosmonaut is vomiting. It is believed he is declaring an emergency and requesting an immediate landing. The State Commission meets again and has to decide within 40 min whether to begin

---

[1] http://www.astronautix.com/k/kamanindiaries.html.

setting the spacecraft up for retrofire. But then when Korolev and Smirnov ask the cosmonaut to verify, he explains "I am excellent, I was observing *meteorological* thunderstorms and lightning". However Gagarin and Kamanin are suspicious of the explanation—they believe Popovich had an attack of nausea, panicked, made the emergency radio transmission, but then felt better and didn't want to admit to his weakness when confronted by the leadership. However it is now too late. He is set to return at nearly the same time as Nikolayev on Vostok 3. Both spacecraft land successfully six minutes apart a short distance from each other. However flight plans for the State Commission are wrecked due to bad weather at nearby airfields.

1962 August 16—Vostok 3/4 post-flight debriefings.

Nikolayev and Popovich finally arrive in Kuibyshev aboard an Il-18 aircraft that originated from from Sary Shagan. Now come the medical check-ups and interviews by the State Commission, The State Commission finds that both missions have outstanding results. The cosmonauts present believe that in the future men, not machines, should pilot the spacecraft. The way was clear for 5–10 Vostok flights in the next year.

Nikolayev's post-flight debriefing: The rocket vibration was not great initially, but very forceful at the end of operation of the second stage. There was quite a shock on separation of the spacecraft from the third stage. 15 min before the launch of Popovich's spacecraft I oriented the Vostok and at 11:03 the spacecraft was at the correct 73° pitch attitude. However I was unable to see either Popovich's spacecraft or his booster rocket. I had bad communications with Zarya on the first day. On the fourth revolution, during the communications session with Khrushchev, I could not hear, but then during the second, third, and fourth day of the flight communications were clear. The Globus instrument was valuable. Zero-G was not unpleasant, and on the fourth day I sharply turned by head to the left and right but could not force any bad reactions. I felt fully trained in use of the equipment. Over Turkey I could see airfields, cities, paved roads, and ships at sea. The TDU retrorocket operated for 42 s. The re-entry capsule revolved randomly on reaching the denser atmosphere and I pulled 8–9 G's on re-entry. There were many boulders in the landing area, but I was able to guide my parachute to land in a 2 × 2 m clear area.

Popovich debriefing: I could easily see the earth flowing below. Manual orientation using this by day or the stars by night was possible. There was lots of static on the UHF band on space-ground communications. Space-to-space communications with Sokol were very good, especially over the equator. Moving my head caused no motion sickness problems. After ejection, I secured my reserve parachute (as had Nikolayev). I saw a search aircraft twenty minutes after landing. The NAZ antenna did not deploy (as with Nikolyaev).

After the debriefing, a celebration is held with the cosmonauts, State Commission, and local officials. Everyone gets pretty drunk. Kamanin is finally instructed to take Nikolayev and Popovich to bed at midnight. The rest continue until 2 in the morning.

1962 August 17—Vostok 3/4 post-flight.

The cosmonauts continue their post-flight medical examinations, but everyone is suffering from hangovers from the celebration the night before. There was a stupid incident, with some of the leaders blaming Nikolayev of bad behaviour. Most of the commission leaves in the evening. In the afternoon the new heroes of the cosmos—Gagarin, Titov, Nikolayev, and Popovich—are taken boating, to the acclaim of crowds on the shore.

1962 August 22—Future Vostok flight plans discussed.

At Baikonur for the launch of a Venera probe, the Soviet space leadership discussed future plans. The female cosmonaut training group was there for their first rocket launch. The next Vostok would carry the first woman into space; Ponomaryova, Solovyova, and Tereshkova were the leading candidates. Flight plans were discussed at a meeting in the evening between Kamanin and Leonid Smirnov. It would be possible to make the flight by the end of 1962, but March–April 1963 was more likely, depending on the final report on the Vostok 3/4 flights. The work force would be fully occupied in August-October in launching probes to Venus and Mars, also probably delaying any Vostok flight until the following spring. The next flight would probably be part of a group flight of two or three spacecraft, piloted by both men and women. The female flights would be limited to three days, while the male flights would last for 7–8 days.

Although Smirnov spoke of up to five Vostok flights in 1963, there were actually only two complete Vostok spacecraft left. Korolev still claimed the first unpiloted Soyuz test flight could take place in May 1963. The Mars and Venus probes didn't bring any military and very little propaganda advantage to the Soviet Union, in the opinion of Kamanin. He wished that instead Korolev would use those resources for further manned flights, including orbital stations and moon landings. On the other hand the military leadership was even opposed to the modest existing manned space programme. Malinovskiy had blocked attempts to authorise a further ten Vostoks a year earlier. Korolev, Keldysh, and Smirnov were discussing sending a letter directly to Khrushchev, bypassing the General Staff, to plead for more support for manned space flight.

1962 August 24—Baikonur conditions.

Kamanin is at Tyuratam for the impending Venera launch, together with some of the cosmonauts. He notes that officers at Tyuratam have to live in hostels, without their families. Some have been there from three to five years, separated from their wives and children. Those who leave to see their families are court-martialled for desertion. At a morning briefing a new 'forced' method of manually orienting the Vostok is discussed. This will allow the spacecraft to turn 360° in 12 min. The conservative method using residual angular velocities takes two hours. In the evening the State Commission for the Venera launch meets. This is the first one ever not attended by Korolev—after the meeting in the Kremlin, he became very ill, and is in the hospital. It will be two to three weeks before he can return to work.

1962 August 27—Female Vostok flights delayed to 1963.

The prospects did not look good for authorisation of production of ten further Vostok spacecraft. In a heated discussion between Rudenko, Ivanovskiy, and Grechko, it was argued that production of further Vostoks would delay flight of the first Soyuz spacecraft by a year. On the other hand this would mean no Soviet manned flights in 1963–1964. Furthermore Ivanovskiy reported that production of the female version of the Vostok space suit could not be completed until the end of 1962. Therefore this meant that the flight of two female cosmonauts in the final two available Vostok spacecraft would be delayed until March–April 1963—the very end of the storage life of the spacecraft.

1962 September 13—General Staff tries to prevent further Soviet manned spaceflights.

At a meeting of the General Staff on space plans, it was reported that the Ministry of Defence supported completion of two additional Vostok spacecraft to allow four Vostok flights in 1963. But Malinovskiy was adamant: the Vostok fullfilled no military objectives, would not be accepted for military use, and he would recommend to the Military Industrial Commission that the additional flights be rejected. Kamanin noted that history was repeating itself—fifty years earlier Tsarist generals had rejected the acquisition of aircraft by the Imperial Russian Army.

1962 November 9—Plans for additional Vostoks quashed.

Kamanin prepared recommendations for General Staff discussions on future Vostok military flights. His plan involved construction of ten additional spacecraft including new versions to test military equipment for reconnaisance, interception, and combat objectives. Flights would begin in 1963: manned flights of ten days duration; flights with biological payloads of 30 days duration; flights with biological payloads in high orbits to test the effects of Van Allen radiation belt exposure; flights that would conduct a range of technology experiments, including manual landing; landing with the cosmonaut within the capsule; depressurisation of the capsule to vacuum test equipment and suits for future spacewalks; etc.). The plan was killed by his superiors.

1962 November 16—Meeting of the Soviet Ministers.

They agree to a plan for a national centrifuge facility: specifications to be determined in 1963, and the facility completed by 1967. They are not if favour of building more Vostoks—they want to move on to the Soyuz spacecraft. But this will produce an 18–24 month gap in Soviet manned spaceflight, during which the Americans will certainly catch up (Cooper's one-day Mercury flight is already scheduled).

1962 November 29—Final tests for female cosmonauts.

Academic examinations were completed of the female cosmonaut corps. Kuznetsova had missed to much training and was excluded from even taking the test. Of the four women remaining, only Tereshkova did not receive the highest marks. This was attributed to her being too nervous and excited during the examination. All were given the rank of Junior Lieutenant in the VVS Soviet Air Force.

Kamanin considered Tereshkova as the leading candidate for the first flight, with Solovyova as her back-up. In personality they were equivalent to Gagarin/Nikolayev—indeed, Tereshkova was considered 'Gagarin in a skirt'. Ponomaryova and Yerkina were equal candidates for the second female Vostok flight. The group would go to a resort in the Urals from 30 November to 10

January. The final decision as to which one would fly would only be made 3 or 4 days before the flight.

1962 December 27—Absurd situations!

A decree ordering the training of sixty cosmonauts has been laying around, and suddenly the leadership wants to enforce it. 15 new trainee male cosmonauts, and 15 women are to be recruited—an overall total of 20 by the end of 1962 and 40 by the end of 1963 And crews are to be formed and trained, even though there are no spacecraft being built for the missions. And the decision that Popovich is to go on his Cuba tour is handed down only 2.5 h before he is supposed to depart.

1963 January 7—Seven Vostok flights planned in 1963.

Agreement was finally reached among space management for the production of five additional Vostok spacecraft during 1963. Two would be used in solo flights and five in group flights.

1963 January 11—Korolev lays out detailed plan for future Vostok flights.

Korolev and Kamanin meet to lay out Vostok flight plan. There were three variants possible for the March flights: (1) A single female flight of 2–3 days; (2) Two female flights launched one day apart, but landing at the same time; (3) An 'absurd' version: launch of a female cosmonaut for a three day flight, followed two days after her landing by a male cosmonaut on a 5–7 day flight. The planners selected the two female flight variant.

1963 January 17—Cosmonaut PR training.

The cosmonauts need to be trained for press conferences. Nikolayev is to receive special training, as well as Popovich who is being criticised for mistakes made during his Cuba tour. He told reporters 'We will assist Cuba not just on the earth, but from space', and 'The world will soon learn the names of all of the first cosmonaut team', neither of which are state policy.

1963 January 18—Soyuz expert commission.

Smirnov insisted on the following after reviewing Korolev's design: (1) there must be a space suit for every crew member; (2) the spacecraft must be able to use lift during re-entry to change its landing point; (3) the spacecraft must have ejection seats. Korolev and his assistants categorically rejected these demands. Smirnov was only insisting on the availability of suits, not that they be worn at all times; and only on small lifting surfaces to give the capsule more manoeuvrability during re-entry. But Korolev rejected even this. Later the commission went to Chelomei's bureau to see his Raketoplan manned spaceplane design. But this was not even laid out on paper yet, with the draft project not scheduled to be completed until the end of February. Chelomei has already been working on this for two years. In January 1961 he gave a presentation to the General Staff and made big promises in regard to this spacecraft—but nothing has been completed. The only spacecraft that will be realistically available in the next three to five years is Korolev's—anything else would only be purely experimental.

1963 January 31—Smirnov opposed to dual female Vostok flight.

Smirnov only wants to fly two, not four Vostoks this year. One male, and one female cosmonaut would be launched in a group flight. Correct approvals cannot be obtained in time for manufacture of four Vostoks until August of this year. Later Kamanin has another scene with Titov. The cosmonaut was drunk on a factory visit, and defied the militia when confronted.

1963 February 16—Plethora of projects.

Vershinin says the Soviet Union can't work on the Vostok, Soyuz, and Raketoplan manned spacecraft all at the same time. But he still wants to fly four Vostoks by the end of the year.

1963 February 18—Soviet Ministers' decree on use of Vostok.

The Soviet Ministers finally issued decree 24. Four additional spacecraft are to be completed in the first half of 1963. Together with the two existing spacecraft, these will be used for two female flights, three male flights of up to ten days duration, and one 30-day biosat flight.

1963 March 2—Plan for Cosmonaut Training.

The big question regards Gagarin. Shall the 'Columbus of the Cosmos' be allowed to risk his life on another spaceflight? Most of the Soviet leadership are against it, but Gagarin himself wants to train and fly again. Later in the day the cosmonauts have an idiotic argument with IAKM on high-G centrifuge runs for female cosmonauts. This is the first cosmonaut revolt against the policies and practices of IAKM.

1963 March 21—Vostok programme cut back—second female flight cancelled.

Vostok flight plans were drastically curtailed at a meeting of the Presidium of the Communist Party. Korolev presented the plan for 1963 as approved by the Interorganizational Soviet at the beginning of the year. This plan, already in an advanced stage of execution, was rejected utterly by Kozlov and Vershinin. The Ministry of Defence announced its categorical opposition to further Vostok production. It was finally decided that there would be only two flights in 1963 using existing spacecraft. These were scheduled for June and would consist of simultaneous female and male flights. Kamanin was infuriated that although he was ordered by a leadership decree in December 1961 to train five women for spaceflight, the same leadership was now asking—Who ordered this? What was the purpose? Are we sure they're ready?

1963 March 24—VVS Chief of Staff Malinovskiy says that manned flights should be cut back due to safety considerations.

Kamanin considers this a strange attitude—many die every day in auto or aircraft accidents, but not one death will be tolerated in the conquest of space. The whole plan for the next Vostok missions are thrown back for reconsideration. Many meetings occur over the next week—the basic question, was the MO/RVSN/VVS interested in manned space flight or not? Finally the decision was made to continue—a 180° reversal from the original position.

1963 March 27—Cosmonauts Nelyubov, Anikeyev and Filatyev were arrested drunk and disorderly by the militia at Chkalovskiy station.

This was not the first time. The VVS hierarchy wants them all dismissed from the cosmonaut corps. Gagarin says that only Filatyev should be fired. Kamanin would prefer to see all three go, but cannot afford to lose 25% of his flight-ready cosmonauts. He would hope to at least keep Nelyubov, who was a candidate for the third or fourth Vostok flights, but did not perform well on the centrifuge.

1963 April 6—The General Staff considers the topic of spaceflight and is opposed to greater VVS participation.

They are not against the flight of four Vostoks in 1963, though.

1963 April 13—At a meeting with the VVS, Korolev outlines his revised plans for the next fights.

He plans a male flight for 8 days, during which a woman would be sent aloft for 2–3 days.

1963 April 13—Decree issued for four Vostok flights in 1963.

Decree issued by the Soviet ministers and Central Committee setting out four Vostok flights in 1963. Two are to be launched by 15 June.

1963 April 17—Nelyubov, Anikeyev, and Filatyev dismissed from the cosmonaut corps.

The VVS General Staff issues a decree discharging Nelyubov, Anikeyev, and Filatyev from the cosmonaut corps.

1963 April 28—N1 Plans.

An Inter-Institution Soviet considers Korolev's N1 plans. He believes the first booster will be launched in 1965. The N1 is to have a payload capability of 75 tonnes to a 250 km altitude orbit, 50 tonnes to a 3000 km altitude orbit, and 16 tonnes in geostationary orbit. It could launch spacecraft capable of landing men on the moon and returning them to earth, or manned flybys of Mars or Venus. Three to ten launches would be needed for such missions, with the components being docked together in low earth orbit. The N1 can also be used to launch a large space station for military research. After the N1 discussion a decision is made that cosmonauts will not have to spend more than three to four days in a spacecraft mock-up on the ground to prove their readiness for flight. A simulation of the entire flight duration is not necessary.

1963 April 29—IAKM Meeting.

In a bitter and exhausting meeting with IAKM, the decision that cosmonauts will not be required to spend more than 3–4 days in a trainer on earth qualifying for a mission is confirmed.

1963 May 4—Kamanin informed that a dual spaceflight has been decreed within the next 6 weeks.

Only today is Kamanin informed that a dual flight has been decreed within the next 3–6 weeks. The women are ready, but Bykovskiy and Volynov need a few parachute jumps and training in the hot mock-up. Leonov and Khrunov need additional centrifuge training as well. Bykovskiy and Volynov should be ready by 30 May, and Leonov and Khrunov by 15 June. Therefore earliest possible launch date is 5–15 June. Alekseyev's bureau is as always the pacing factor. He can adapt one of the female ejection seats for Bykovskiy, but not for Volynov. The space suit for Leonov will only be completed by 30 May. Kamanin talks to Korolev about dumping Alekseyev's bureau in the future. Cosmonaut parachute trainer Nikitin agrees that Bykovskiy can complete his parachute qualification at Fedosiya on 9–10 May. Further bad behaviour by Titov is reported during a trip to Kiev. He insulted an officer ('I am Titov, who are you?') and then had general's wives intervene on his behalf to get him out of trouble.

1963 May 7—Yerkina excluded from Vostok 6.

Yerkina was excluded from Vostok 6 due to her performance during the three day test in the hot mock-up. She took off her boots after one day, and ate only three rations in three days. She was weak and fainted after coming out of the spacecraft.

1963 May 11—Vostok 5/Vostok 6 Planning.

Korolev reports still problems with components of the electrical system from the Kharkov factory—the same problems that existed in 1962. The cosmonauts will go to Tyuratam on 27/28 May, with launch planned for 3/5 June. Bykovskiy is named prime for Vostok 5, with Volynov his backup. Tereshkova is named prime for Vostok 6, with Solovyova and Ponomaryeva both as her backups. This selection is however made despite strong support for Ponomaryeva as prime by Keldysh and Rudenko.

1963 May 13—Korolev fights excessive VVS staff at Tyuratam.

The VVS wants to send 55 staff to Tyuratam for the launches, but Korolev wants no more than 25. This is just possible—11 cosmonauts, 8 engineers, and vital support staff only. Bykovskiy was to start a two day run in the hot mock-up, but it was called off due to defects with his suits—the biosensors were wired to his helmet microphone! The suit seems not even to have been tested before delivery. Alekseyev was supposed to have it ready by 9 May, now it will only be ready for use by 14 May. Gordon Cooper is scheduled for a 34 h Mercury flight tomorrow....

1963 May 14—Tereshkova and Solovyova rated most ready to fly on Vostok 6.

Tereshkova and Solovyova are most ready to fly and will be sent to Fedosiya for sea training first. Ponomaryova and Yerkina will follow tomorrow. Bykovskiy started his run in the hot mock-up at 10:00 am.

1963 May 15—Cooper's flight scrubbed; Bukovskiy to start in Vostok 5 hot mock-up.

Cooper's flight was scrubbed due to a problem with the Bermuda tracking site. Bykovskiy's suit microphone failed on the second day in the hot-mock-up and he as to communicate by telephone or telegraph. The doctor's insistence that each cosmonaut spend the full duration of his planned flight in the hot mock-up is idiotic. The US practice is to simulate the active portions of the flight only. In actuality every day spent in a suit on the earth is as gruelling as three days in space.

1963 May 16—Bykovsky's ordeal in Vostok-5 hot mock-up to be ended on third day.

It is decided that extending Bykovskiy's ordeal in the hot mock-up to a third day makes no sense. The IAKM doctors are utterly incompetent. Cooper has landed after a successful flight. The US is now hot on our tail in the space race.

1963 May 17—Problems with Titov again.

Problems with Titov again. While on a road trip with a journalist, he left a satchel with sensitive and classified papers unattended in his car—documents from Korolev, secret state decrees by the Supreme Soviet, etc. At 12:30 Volynov took Bykovskiy's place in the hot mock-up. Examination of Bykovskiy's suit showed that it had been incorrectly assembled.

1963 May 21—The cosmonauts are informed of the selections for the Vostok 5/6 flights.

Korolev asks Ponomaryova why she is so sad—'I am not sad, but serious, as always'.

1963 May 28—Cosmonaut's parachute trainer Nikitin killed in an accident.

He tangled in the air with another member of a group jump, Aleksei Novikov. Both were killed. The Vostok 5 and 6 launch vehicles and spacecraft are both in the MIK assembly wall. Work began on them two weeks ago. Nevertheless Korolev is not happy with the results. He wants the tests run over from the start. Round-the-clock work begins from this day. The bad weather and the news of Nikitin's death produce an atmosphere of gloom. Nikitin's funeral is scheduled for 30 May. Therefore the cosmonauts have delayed their departure in order to attend the funeral and will not arrive at Tyuratam until 31 May. Kamanin was very worried about the effect of Nikitin's death on the female cosmonauts' nerves. The final decree set the launch dates as 2/3 June, with landing on 7/8 June. Kamanin gets into a heated argument with Rudenko, who wants to fly all of the cosmonauts to Tyuratam on a single aircraft. He doesn't see what the big deal is—after all, state ministers fly together all the time.

1963 June 1—Cosmonauts and brass arrive at the cosmodrome for the Vostok 5/6 launch.

A meeting is held to discuss emergency recovery of the Vostoks. There is no realistic chance of their survival if they land at sea in the South Atlantic, Pacific, or Antarctic Oceans, however plans must be made. Several ships and three to four Tu-114 aircraft would be required to have any realistic chance of recovery. However these are not available.

1963 June 3—Vostok 5/6 Flight Preparations.

At 9 am Tereshkova, Solovyova, and Ponomaryova practice donning and doffing their space suits. Bykovskiy and Volynov prepare their ship's logs. Korolev discusses plans for tests of the cosmonaut's ability to discern objects from space. Colonel Kirillov completes preparation of the spacecraft for flight.

1963 June 4—The State Commission for Vostok 5/6 launches meets.

All is ready, but the wind is predicted to by 15–20 m/s on 7 June. The launch vehicle cannot be launched in winds over 15 m/s. Bykovskiy and Tereshkova are confirmed as the crew for 8 and 3 day flight durations. When they return to earth, a new and difficult life as celebrities will begin for them—they will be known all over the world.

1963 June 5—Vostok 5/6 Flight Preparations.

On the last five days it has been 25 °C during the days and 15 °C at night. In the evening the classified film on Nikolayev and Popovich's flights is screened. Kamanin regrets that it cannot be made public. What the Soviet state considers secrets—the configuration of the rocket and spacecraft, the identity of the managers and launch teams—are public knowledge in the US program. A VVS Li-2 (DC-3) transport arrives at Tyuratam with three tonnes of fruit. A real treat for the launch teams. The cosmonauts spend their final night in the cottages. These are equipped with good-quality Italian air conditioners that keep the cosmonauts comfortable on their last night on earth.

1963 June 6—Launches of Vostok 5 and 6 delayed.

Launches of Vostok 5 and 6 are delayed due to failure of the command radio line. There were many such failures during preparation of the spacecraft. It will take three to four days to fix. Kamanin inspects the site for the planned cosmonaut quarters on the Syr Darya river. It is located next to Khrushchev's houses (which he has handed over to Chelomei for quartering his people) and the television centre. The building will face east, with a view of the river and a wooded island. Bykovskiy is run through a first 'practice press conference' to teach him the correct responses to questions. The military officers want to minimise press contacts with the cosmonauts in any case. But the kids in the town are mad about the cosmonauts—the chanted from 6 to 11 pm in the evening outside their quarters, and Kamanin has seen teenage girls stand in the rain for hours for a chance to see Titov (and he never even came out as promised).

1963 June 8—Vostok 5/6 Flight Preparations.

A review of the spacecraft radio problems shows that the rejection rate for production equipment is 6% against 2% guaranteed by 5-GURVO.

Tereshkova sits in the Vostok 6 spacecraft, and makes a good impression on the technicians.

1963 June 9—Vostok 5 is rolled out.

Vostok 5 is rolled out to the pad at 9 am. It is erected and then tested from 11:00 to 13:30. All is well and it is declared ready for launch. At 16:00 the cosmonauts take the traditional pre-launch walk along the Syr Darya. All is filmed for posterity, including the cosmonauts fishing for their dinner.

1963 June 10—Vostok 5 scrubbed due to solar flares.

The launch of Vostok 5 is set for 11 June. Final training and consultations are under way. Korolev is not happy with the condition of the spacecraft. At 22:30 in the evening the launch is scrubbed when Keldysh calls from Moscow and advises excessive solar flare activity is expected. Keldysh will review the data tomorrow and advise if it really poses a danger to the cosmonauts.

1963 June 11—Vostok 5 slipped to 14 June.

The cosmonauts spend the day on the beach. Tereshkova sits a long time with Korolev on the balcony on the second floor of the house on the river. He interviews here thoroughly to make sure she is ready for the flight. The State Commission meets at 17:00. The expected solar flare did not occur, but the Crimean Observatory claims the risk will remain high. The decision is made to defer the launches to 14/15 June.

1963 June 12—Vostok 5 preparations.

The next two days are spent waiting—on the beach in the heat, in fishing, and in politics between the brass at the site.

1963 June 13—Vostok 5 a go for 14 June.

The solar activity has subsided and the launch of Vostok 5 is set for the following day. Kamanin has foreboding about the flight—eight days in space will be tough on both man and machine.

1963 June 14—Vostok 5 Launch.

At 8 am the State Commission meets and approves a five-hour countdown to launch of Vostok 5 at 14:00. The cosmonaut and his backup have slept well and are at medical at 9:00 for the pre-flight physical examination and donning of their space suits. At T minus 2 h and fifteen minutes they ride the bus to the pad. A few minutes after Bykovskiy is inserted into the capsule, problems with the UHF communications channels are encountered—three of the six channels seem to be inoperable. Gagarin and Odintsov are consulted on how it will be for the cosmonaut to fly with just three channels operable—is it a Go or No-Go? Go! Next a problem develops with the ejection seat. After the hatch is sealed, a technician cannot find one of the covers that should have been removed from the ejection seat mechanism. It is necessary to unbolt the hatch and check—the seat will not eject if the cover has been left in place. At T minus 15 min Gagarin, Korolev, Kirillov, and Kamanin go into the bunker adjacent to the rocket.

A new problem arises—the 'Go' light for the Block-E third stage won't illuminate on the control room console. It can't be determined if it is a failure of the stage or an instrumentation failure. It will take two to five hours to bring up the service tower and check out the stage. But if the rocket is left fuelled that long, regulations say it must be removed from the pad and sent back to the factory for refurbishment. In that case there can be no launch until August. Krylov and the State Commission would rather defer the launch to August. The last possible launch time is 17:00 in order to have correct lighting conditions for retrofire and at emergency landing zones. But Korolev, Tyulin, Kirillov, and Pilyugin have faith in their rocket, decide that the problem must be instrumentation, and recycle the count for a 17:00 launch.

The launch goes ahead perfectly at 17:00—even all six UHF communications channels function perfectly. On orbit 4 Bykovskiy talks to Khrushchev from orbit and good television images are received from the capsule. Bykovskiy reports he can see the stars but not the solar corona. His orbit is good for eleven days.

1963 June 14—Tereshkova meets with the command staff at 17:00, followed by dinner.

Tereshkova meets with the command staff at 17:00, followed by dinner. She has a good appetite and is ready to go for her space flight.

1963 June 16—Vostok 6.

Joint flight with Vostok 5. First woman in space, and the only Russian woman to go into space until Svetlana Savitskaya 19 years later. On its first orbit, *Vostok 6* came within about five km of *Vostok 5*, the closest distance achieved during the flight, and established radio contact. Flight objectives included: Comparative analysis of the effect of various space-flight factors on the male and female organisms; medico-biological research; further elaboration and improvement of spaceship systems under conditions of joint flight. It was Korolev's idea just after Gagarin's flight to put a woman into space as yet another novelty. Khrushchev made the final crew selection. Korolev was unhappy with Tereshkova's performance in orbit and she was not permitted to take manual control of the spacecraft as had been planned.

1963 June 16—Vostok 5 day 3/Vostok 6 launch.

Bykovskiy slept well, his pulse was 54. The ground station could observe him via television—he made no motion while sleeping. On orbit 23 the cosmonaut was to communicate with earth, but no transmissions were received. Gagarin asks him why, and Bykovskiy simply replies that he had nothing to say and had already had a communications session with Zarya-1. But this was not true, they also reported no transmissions. At 07:00 he is asleep again, pulse 48–51. An hour later Korolev calls and discusses the impending launch of Vostok 6, 11 h later.

At 12:15 Tereshkova is on the pad. Her pulse skyrockets to 140 aboard the elevator to the top of the rocket. 10–15 min later she is in the capsule and testing radio communications with ground control. There are no problems with the spacecraft or launch vehicle during the countdown—everything goes perfectly, just as it did on 12 April 1961 when Yuri Gagarin became the first man in space. Tereshkova handles the launch and ascent to orbit much better than Popovich or Nikolayev according to her biomedical readings and callouts. Kamanin feels reassured that it was no mistake to select her for the flight.

The launch of the first woman into space creates a newspaper sensation throughout the world. Direct orbit-to-orbit communications between Tereshkova and Bykovskiy are excellent. She talks to Khrushchev and the Soviet leadership soon thereafter. This was truly a great victory for Communism!

1963 June 19—Vostok 5 and Vostok 6 return to earth.

In the morning Tereshkova manually oriented the spacecraft for re-entry easily and held the position for 15 min. She was very happy with the result. At 9:00 the state commission took their places in the command post. At 9:34:40 the retrofire command was sent to Vostok 6. After a few seconds, telemetry was received indicating that the engine burn was proceeding normally. The nerves of the commission members finally settled down, but Tereshkova did not call out each event as required. No report of successful solar orientation was received, no report of retrofire, and no report of jettison of the service module. Things remained very tense in the command post—no communications were received from the capsule at all. Knowledge that the spacecraft was returning normally were only received via telemetry, including the signal that the parachute opened correctly from above the landing site. Both spacecraft landed two degrees of latitude north of the aim point. It was calculated that this could have occurred by duplicate landing commands having been sent, but such a failure could not be duplicated in post-flight tests of ground equipment.

Many errors occurred in the entire landing sequences, including actions of the VVS recovery forces. The conditions of the cosmonauts were only reported several hours after their landings. Big crowds gathered at both landing sites. Bykovskiy spent the night in Kustan, then left on 20 June aboard an Il-14 for Kuibyshev. Tereshkova spent her first night in Karaganda, then flew in an Il-8 to Kuibyshev. Many congratulatory phone calls were received from the Soviet leadership. Korolev declared he had no longer had the time to personally direct Vostok flights and wanted to hand the spacecraft over to the military for operational use. He could then concentrate on development of the Soyuz and Lunik spacecraft.

1963 June 20—Vostok 5/6 cosmonaut debriefing.

Korolev, Tyulin, and Rudenko left Tyuratam aboard an An-12, followed by 60 others (cosmonauts, officers, engineers) aboard an An-10. General Goreglyad requests that 'extraneous' staff remain in Kuibyshev, while the rest will proceed on to Moscow with Bykovskiy and Tereshkova. The aircraft arrive at 11:30 in Kuibyshev, then go to the debriefing building on the Volga river. There the debriefing of the two cosmonauts began at 13:00. After the debriefings, in the evening, Korolev took the cosmonauts for a trip on the Volga. Kamanin was infuriated—partying would ruin the post-flight medical tracking.

1963 June 21—Vostok 5/6 cosmonaut debriefing.

Tomorrow morning the entire entourage would depart for Moscow. But on this day at the house on the Volga the cosmonauts were subjected to the attentions of seventy doctors, 100 correspondents, and a large additional number of KGB supervisors, military officers, and engineers. Tereshkova looked fresh and her first press conference with sixty correspondents went well—she made no big errors.

1963 June 22—Vostok 5/6 cosmonaut welcome in Moscow.

The big day for the cosmonauts. Departure for Moscow was scheduled for 10:30, with the meeting with Khrushchev at Vnukovo planned for 15:00. A sensitive issue—who would exit the aircraft first—Tereshkova, the main celebrity, or Bykovskiy, the senior cosmonaut and the first one launched? An enormous motorcade takes the entourage from the house on the Volga to the airport. Tereshkova and Kamanin are in the lead automobile, followed by Bykovskiy in the second, then the correspondents and so far in others, at five minute intervals. Huge crowds all along the route chant 'Valya! Valya! During the flight to Moscow Kamanin goes over Tereshkova's speech with her. When she and Bykovskiy get off the plane and march up to the tribune, a completely new life will begin for them. After the immense reception at the airport, they go with the leadership to a huge rally at Red Square.

1963 June 24—Controversy over Tereshkova's performance.

The cosmonauts are prepared by Keldysh, Tyulin, and Korolev for their first big press conference. Yazdovskiy has inserted a paragraph in the official press release about Tereshkova's poor emotional state while in space. He claims she experienced overwhelming emotions, tiredness, and a sharply reduced ability to work and complete all of her assigned tasks. Kamanin takes him aside and asks him not to exaggerate her difficulties during the flight. She only had tasks assigned for the first day. When the flight was extended for a second, and then a third day, there was essentially nothing for her to do. The ground command did nothing to support her during those additional days. She certainly was never tired, never objected, but rather did all she could to complete fully the flight program.

1963 June 25—Vostok 5/6 returned cosmonauts traditional meeting with Korolev.

The returned cosmonauts have the traditional meeting with Korolev at the design bureau and hand over their flight logs. The new cosmonaut group is presented as well. Korolev is in a good mood, and makes an especially long-winded speech. Tereshkova has to leave early, at 12:00, to attend yet another press conference and a woman's congress. These activities kept her going until 22:00 in the evening—a gruelling schedule indicative of what was to come.

1963 June 27—Vostok 5/6 cosmonauts pose for their official colour photographs.

1963 June 29—Vostok 5/6 cosmonauts prepared for first international press conference.

At a meeting of the Central Committee, Tereshkova and Bykovskiy are taken through possible questions and correct replies by Serbin and Keldysh in preparation for their first international press conference. The training extends form 10 in the morning to 17:00 in the afternoon.

1963 July 1—Vostok 5/6 international press conference.

Big international press conference with the cosmonauts, beginning at 13:00. The session goes 1 h and 45 min and all answers given by the cosmonauts are acceptable. After this conference they disappear from public view for seven days of medical examinations and monitoring.

1963 July 2—Yazdovskiy presses complaints about Tereshkova's performance.

Doctor Yazdovskiy is insisting that Tereshkova is not being truthful about her flight experience. She handed out rations to on-lookers at the landing site in order to cover up the fact she did not eat enough during the flight. Kamanin considers the accusation a stupidity and indicative of the constant war going on between the flight surgeons and the cosmonauts. Tereshkova powerfully denies the accusation and defends herself well.

1963 July 3—Cosmonaut controversies.

A fight ensues over the release of the motion picture film of the flight. The Kremlin leadership still does not want to show the 'secret' launch cadres, rocket and spacecraft configurations, etc. There is also conflict with the planned dismissal of cosmonauts Nelyubov, Anikeyev, and Filatyev, with the flown cosmonauts using their connections with the political hierarchy to try and overturn the decisions of their military commanders. Finally, Tereshkova started a campaign to get a posthumous Hero of the Soviet Union medal for cosmonaut parachute trainer Nikitin. This particularly irritated the military command since as far as they were concerned Nikitin died due to his own error and killed another parachutist in the process. In no way was this deserving of a medal, but the award would convey significant financial benefits to his family and Tereshkova fought on. This was indicative of the quick turnaround celebrity brought to the cosmonauts—from obedient junior officers, anxious not to lose a chance for a spaceflight, to aggressive campaigners, willing to take on even members of the General Staff for what they thought was right.

1963 July 7—Kamanin presses for specialised cosmonaut training.

In a two hour meeting with Rudenko, Kamanin attempts to convince him of the need for specialised cosmonaut training (qualifying as spacecraft commander, pilot, navigator, engineer, etc.) for future multi-crew spacecraft. Kamanin points out that in five to seven years they will be routinely flying 2 or 3 place spacecraft and need to start differentiating training now in order to be ready in time. However Rudenko remains unconvinced. Meanwhile Bykovskiy and Tereshkova are at the cosmonaut training centre, completing their flight reports. Kamanin faces difficulties in booking a hotel for the entire cosmonaut group in the Crimea in August—he can't find any place with fifty vacancies, and concludes he'll have to split the group up. Pressure is coming from the Foreign Ministry for Tereshkova to make an early trip to Brazil, but she is already booked for two or three tours of friendly socialist countries beginning on 30 August and any additional trips can only be made after those are completed.

1963 July 10—Odintsev pressing criticism of Tereshkova.

Odintsev is still trying to formally criticise Tereshkova for her flight performance. He charges that she was drunk when she reported to the launch pad and while in orbit was insubordinate, disregarding direct orders from the Centre. Kamanin knows this to be absolutely not true. Both cosmonauts and workers at the cosmonaut training centre report that is impossible to work with Odintsev any more—they want him out.

1963 July 12—Korolev wants review of Tereshkova's flight performance.

Kamanin discusses future cosmonaut book plans with writer Riabchikov. He is interrupted by a call from Korolev. Korolev wants Tereshkova and Bykovskiy in his office the following morning at 10 am sharp and he wants a full explanation for Tereshkova's poor self-samochuviniy on orbits 32 and 42, about her vote, her poor appetite during the flight, and her failure to complete some assigned tasks. He blames Kamanin for providing her with inadequate training prior to the flight—which Kamanin finds a joke since he had never received any support in the past from Korolev for his requests for more and better training of the cosmonauts in high-G and zero-G situations. Korolev had also never listened to any of Kamanin's complaints about the need to improve the living conditions for the cosmonaut on the Vostok spacecraft.

1963 July 13—Bykovskiy and Tereshkova take their first road trip.

Bykovskiy and Tereshkova take their first road trip, to Yaroslavl. It is clear that Tereshkova is the star and Bykovskiy is in her shadow. Bykovskiy calls Kamanin—he asks that his wife and Tereshkova's brother be allowed to accompany them on their first foreign trip. Kamanin rejects the request.

1963 July 17—Rudenko meets Odintsev.

Rudenko meets Odintsev but does not give him the word of his removal directly. The decision will wreck Odintsev's career—his next assignment would have been command of an Air Army. Odintsev fretted over the number of stars on his uniform and fawned over academics—he never looked after his own people, which would have prevented things coming to this.

1963 July 19—Cosmonaut tour plans through December 1963.

Cosmonaut tour plans are firmed up for September-December 1963.. Tereshkova and Bykovskiy are to be given a gruelling schedule, having to visit Bulgaria, Mongolia, Italy, Switzerland, Norway, Mexico, India, Ghana, and Indonesia.

1963 July 22—Conference on space cabin ecology.

Keldysh, Korolev, Voronin, and Kamanin attend a conference on space cabin ecology. Presentations are made by IAKM, OKB-124, the Biology Institute, and the Physiology Institute. In two to three years the USSR expects to orbit spacecraft of 78–80 tonnes, which will be assembled in earth orbit to produce larger spacecraft. These will not only fly around the moon, but also be used to fly to Venus, Mars, and other planets. Although it will take years, many technical problems have to be solved before such a spacecraft can be built. How to shield the crew from radiation? How best to regenerate the air? How to recycle the water? Can the crew survive for long flights in zero-G, or must some form of artificial gravity be provided? If so, what is the best method?

How can the psychological health of the crew best be maintained on long flights?

It is reported that a lot of test stand work has been completed and is underway on closed ecological systems for recycling the air and water. One kilogram of chlorella algae can produce 27 kg of oxygen per day. Since each man will require 25 kg of oxygen per day, 2 kg of chlorella per crew member will be adequate. Therefore the problem of recycling the cabin atmosphere is considered already solved.

Food requirements per crew member are 2.5–3.0 kg/day, or about one tonne per year. It is expected that in two to three years development will be complete of a system that will recycle 80% of the food. A 150 kg device will produce 400–600 g of food per day, or 100–200 kg per year.

1963 August 2—No further Soviet manned flights in 1963.

It is clear that there may be no Soviet manned flights in 1963, and certainly not in the spring. It is possible the unmanned biosat Vostok will be flown in the second half of 1963. Korolev's plate is full with other work—Soyuz development, several Zenit reconnaissance satellite launches, lots of Luniks. Meanwhile Kamanin is completely occupied with cosmonaut tours and publicity. Over 200,000 cosmonaut fan letters have been received—they can't handle them all, a special unit will have to be created just to handle the mail. The KGB has assigned Yevgeniya Pavlovna Kassirova to accompany Tereshkova on her travels. She is a good choice, has foreign travel experience and excellent English.

1963 September 6—Tereshkova accused of a scandal in Gorkiy.

The militia claims that Tereshkova was drunk and created a scandal with a militia officer in Gorkiy. She categorically denies being drunk, but does admit to having a confrontation with a militia captain.

1963 September 7—Tereshkova and Bykovskiy begin an eight day tour of Bulgaria.

1963 September 23—First child born to someone who has been in space.

A daughter is born to Titov. This is the first child born to someone who has been in space.

1963 September 26—Gagarin and Kamanin travel to Paris.

Gagarin and Kamanin travel to Paris. On arrival they are taken to see the Eiffel Tower and a quick tour of the city, which Kamanin finds beautiful.

1963 September 29—Gagarin tours Le Bourget airfield.

1963 September 30—Gagarin visits UNESCO.

Gagarin visits UNESCO in Paris, followed by an interview with Paris Match. In the evening he and Kamanin visit Maxim's. On the same day Tereshkova departs for Cuba from Moscow. This is followed by a visit to Mexico.

1963 October 7—Nikolayev to wed Tereshkova.

Kamanin meets with Nikolayev to discuss the timing for his enforced wedding to Tereshkova. Nikolayev is evasive, doesn't want to set a date, won't give a direct answer. Kamanin points out the wedding will be the subject of a government decree and a precise date must be set. The possible days are limited due to Tereshkova's heavy travel schedule. Nevertheless Nikolayev refuses to commit to a date in October.

1963 October 9—Kamanin and Gagarin fly to Cuba to join Tereshkova.

Kamanin and Gagarin fly to Cuba to join Tereshkova. Then over the next 13 days on to Mexico, USA, Canada, England, and East Germany

1963 October 29—The issue of the Nikolayev/Tereshkova wedding has come to a head.

The wedding has to be arranged with the VVS General Staff in accordance with the resolution of the Central Committee. Kamanin calls Tereshkova and Nikolayev and orders them to decide the issue—the MUST SET A DATE. He is getting ten phone calls a day about it and can resist no longer

1963 October 30—Only at 14:30 due Tereshkova and Nikolayev finally give in.

Only at 14:30 due Tereshkova and Nikolayev finally give in. The wedding is set for three days later. 300 will attend the wedding, from Khrushchev on down.

1963 October 31—The highest leadership of the Soviet Union is busy making the Nikolayev/Tereshkova wedding arrangements.

The highest leadership of the Soviet Union is busy making the Nikolayev/Tereshkova wedding arrangements. Tereshkova disappears for four hours during the day and doesn't show up for important meetings.

1963 November 2—Kamanin's phone is ringing off the hook.

Kamanin's phone is ringing off the hook. Thousands want to attend the wedding—and it turns out there will be only space for 200. Kamanin is taking all the blame for this.

1963 November 3—Nikolayev/Tereshkova wedding.

The wedding, attended by the top leadership of the Soviet Union, goes well. Afterwards the newlyweds continue the party with a friends-only group of 60 at Nikolayev's apartment.

1963 November 5—Khrushchev gives Nikolayev and Tereshkova a new apartment.

Khrushchev has given Nikolayev and Tereshkova a new apartment in Moscow. It is in a building normally reserved exclusively for the highest Communist Part members—Kutuzovskiy Prospect number 30132, Apartment 1013L. The apartment has 7 rooms and can be divided into two sections if they wish to live apart.

Bibliography

AA.VV., *Da Baikonur alle stelle. Il grande gioco spaziale*, Centro Studi Vox Populi (2013)

Martha Ackmann, *The Mercury 13 - The Untold Story of Thirteen American Women and the Dream of Space Flight*, Random House Publishing Group (2003)

Marco Aliberti, Ksenia Lisitsyna, *Russia's Posture in Space*, Springer (2018)

American Astronautical Society, *Space Exploration and Humanity: A Historical Encyclopedia [2 Volumes]*, ABC-CLIO (2010)

Lynne Attwood, Melanie Ilic, Susan E. Reid (edited by), *Women in the Khrushchev Era*, Palgrave Macmillan (2004)

David Baker, *International Space Station - An Insight Into the History, Development, Collaboration, Production and Role of the Permanently Manned Earth-orbiting Complex*, Haynes Publishing UK (2016)

Peter Bond, *The Continuing Story of The International Space Station*, Springer (2002)

Luca Boschini, *Il mistero dei cosmonauti perduti: Leggende, bugie e segreti della cosmonautica sovietica*, CICAP (2013)

Richard Brownell, *Space Exploration*, Greenhaven Publishing LLC (2012)

Sarah Bruhns, *Yuri Gagarin: The Spaceman*, Hyperink (2012)

Colin Burgess, *The First Soviet Cosmonaut Team: Their Lives, Legacy, and Historical Impact*, Springer (2008)

Colin Burgess, *Shattered Dreams - The Lost and Canceled Space Missions*, Nebraska (2019)

Michel Capderou, *Satellites - Orbits and Missions*, Springer (2005)

Giovanni Caprara, *Oltre il Cielo - Incontri straordinari con esploratori della Luna e dello spazio*, Hoepli (2019)

Umberto Cavallaro, *Women Spacefarers - Sixty Different Paths to Space*, Springer (2017)

Boris E. Chertok, *Rockets and People (vol. 1, 2, 3, 4)*, NASA (2005)

Michael D. Cole, *Vostok 1 - First Human in Space*, Enslow Publishers (1995)

Martin J. Collins, Space Race, *The U.S.-U.S.S.R. Competition to Reach the Moon*, Pomegranate Communications (1999)

Luigi T. De Luca, Max Calabro, Toru Shimada, Valery P. Sinditskii (edited by), *Chemical Rocket Propulsion - A Comprehensive Survey of Energetic Materials*, Springer (2016) - ebook

Robert C. Dempsey (Edited by), *The International Space Station - Operating an Outpost in the New Frontier*, NASA (2017)

Steven J. Dick, *Critical issues in the history of spaceflight*, SNOVA (2018)

Steven J. Dick, *Remembering the Space Age*, NASA (2008)

Bartolomeo Di Pinto, Lucia Marinangeli, Enrico Flamini, *Dallo Sputnik a Marte e Oltre*, Youcanprint (2021)

Amalia Ercoli Finzi, *Corsa allo spazio 1, 2, 3... Via!* Dedalo (2021)

Ben Evans, *Foothold in the Heavens - The Seventies*, Springer (2010)

Ben Evans, *Partnership in Space - The Mid to Late Nineties*, Springer New York (2013)

Tim Furniss, David Shayler, Shayler David, Michael D. Shayler, *Praxis Manned Spaceflight Log 1961-2006*, Springer New York (2007)

Jurij Gagarin, *La via del cosmo - Sputnik, Lunik, Vostok: l'assalto sovietico al cielo*, Pigreco (2013)

Jurij Gagarin, *Non c'è nessun dio quassù - L'autobiografia di Gagarin. Il primo uomo a volare nello spazio*, Red Star Press (2014)

Slava Gerovitch, *Soviet Space Mythologies - Public Images, Private Memories, and the Making of a Cultural Identity*, University of Pittsburgh Press (2015)

Slava Gerovitch, *Voices of the Soviet Space Program: Cosmonauts, Soldiers, and Engineers Who Took the USSR Into Space*, Palgrave Macmillan (2014)

Clive Gifford, *The Race to Space - From Sputnik to the Moon Landing and Beyond...*, Words & Pictures (2019)

Michael H. Gorn, *Spacecraft - 100 Iconic Rockets, Shuttles, and Satellites That Put Us in Space*, Voyageur Press (2018)

Umberto Guidoni, *Dallo Sputnik allo Shuttle*, Sellerio Editore (2012)

Rex Hall, David Shayler, *Soyuz - A Universal Spacecraft*, Springer (2003)

Rex Hall, Rex D. Hall, David Shayler, Shayler David, Bert Vis, *Russia's Cosmonauts - Inside the Yuri Gagarin Training Center*, Springer New York (2007)

John Hamilton, *Missiles and Spy Satellites*, Abdo Publishing (2018)

David M. Harland, John E. Catchpole, John Catchpole, *Creating the International Space Station*, Springer (2002)

David M. Harland, *The Story of Space Station Mir*, Springer New York (2007)

James Harford, *Korolev: How One Man Masterminded the Soviet Drive to Beat America to the Moon*, Wiley (1997)

Brian Harvey, *Soviet and Russian Lunar Exploration*, Springer (2007)

Bart Hendrickx, Bert Vis, *Energiya-Buran - The Soviet Space Shuttle*, Springer (2007)

Grujica S. Ivanovich, *Salyut - The First Space Station: Triumph and Tragedy*, Springer New York (2008)

Achille Judica Cordiglia, Giovanni Battista Judica Cordiglia, *Banditi dello spazio. Dossier Sputnik 2*, Minerva Medica (2010)

Andrew L. Jenks, *The Cosmonaut Who Couldn't Stop Smiling: The Life and Legend of Yuri Gagarin*, Cornell University Press (2019)

Monika Koli, *5 Greatest Astronauts of the World*, Prabhat Prakashan (2021a)

Monika Koli, *Top 5 Astronauts of the World*, Prabhat Prakashan (2021b)

Nikita S. Kruscev, *Kruscev Ricorda*, SUGAR EDITORE (1970)

Kirsten Larson, *International Space Station,* Rourke Educational Media (2017)

Allison Lassieur, *International Space Station - An Interactive Space Exploration Adventure*, Capstone (2016)

Roger D. Launius, *Frontiers of Space Exploration*, Greenwood Press (2004)

Alexei Leonov, David Scott, *Two Sides of the Moon - Our Story of the Cold War Space Race*, St. Martin's Publishing Group (2013)

John M. Logsdon, Robert W. Smith, Roger D. Lanius (edited by), *Reconsidering Sputnik - Forty Years Since the Soviet Satellite*, Taylor & Francis (2013)

Paolo Magionami, *Gli anni della Luna - 1950–1972: l'epoca d'oro della corsa allo spazio*, Springer (2009)

Jennifer Mason, *The Space Race*, Gareth Stevens Publishing Lllp (2017)

Paola Messana, *Soviet Communal Living - An Oral History of the Kommunalka*, Palgrave Macmillan (2011)

Ernst Messerschmid, Reinhold Bertrand, *Space Stations - Systems and Utilization*, Springer (1999

Robin Milner-Gulland, *Patterns of Russia - History, Culture, and Spaces*, Reaktion Books (2020)

James Oberg, *Star-Crossed Orbits: Inside The US-Russian Space Alliance*, McGraw-Hill Education (2001)

Lina Orlovsky, David Herzog, *From Sputnik to "Buran"*, eBook GmbH (2015)

John O'Sullivan, *Japanese Missions to the International Space Station - Hope from the East*, Springer (2019)

Ruth Owen, *Spacecraft*, Rosen Publishing Group (2014)

Valeria Palumbo, *L'epopea delle lunatiche - Storie di astronome ribelli*, Hoepli (2018)

Dominic Phelan (edited by), *Cold War Space Sleuths - The Untold Secrets of the Soviet Space Program*, Springer (2012)

Vladimir Propp, *Morfologia della fiaba (1928)*, Einaudi (2000)

Eugen Reichl, *The Soviet Space Program - The N1, the Soviet Moon Rocket*, Schiffer Publishing

Andrey Sakharov, *Memoirs (1992)*, Vintage Books (2011)

David J. Shayler, *Disasters and Accidents in Manned Spaceflight*, Springer (2000)

David Shayler, *Walking in Space*, Springer (204)

David Shayler, Shayler David, Ian A. Moule, *Women in Space - Following Valentina*, Springer (2006)

Daniel H. Shubin, **Konstantin Eduardovich Tsiolkovsky - The Pioneering Rocket Scientist and His Cosmic Philosophy**, Algora Publishing (2016)

Asif A. Siddiqi, *Challenge to Apollo: The Soviet Union and the Space Race, 1945–1974*, NASA (2000)

Asif A. Siddiqi, James T. Andrews (edited by), *Into the Cosmos: Space Exploration and Soviet Culture*, University of Pittsburgh Press (2011)

Michael G. Smith, *Rockets and Revolution - A Cultural History of Early Spaceflight*, Nebraska (2014)

Soviet Space Programs 1976–80: Unmanned space activities, U.S. Government Printing Office (1982)

Soviet Space Programs 1981–87 : Piloted Space Activities, Launch Vehicles, Launch Sites, and Tracking Support, U.S. Government Printing Office (1988)

Giles Sparrow, Judith John, *Exploring Space*, Cavendish Square (2015)

Giles Sparrow, Judith John, Chris McNab, *Shuttles and Space Missions*, Cavendish Square Publishing LLC (2015)

Vasil Teigens, Peter Skalfist, Daniel Mikelsten, *La conquista dello spazio*, Cambridge Stanford Books

Colin Turbett, *Soviets in Space - The People of the USSR and the Race to the Moon*, Pen and Sword History (2022)

Ilja Utekhin, *Ocherki kommunalnogo byta*, O.G.I. (2004)

Douglas A. Vakoch, *On Orbit and Beyond - Psychological Perspectives on Human Spaceflight*, Springer Berlin Heidelberg (2012)

Carl von Clausewitz, *On War (Vol. 1, 2, 3)*, David Campbell (1993)

Manfred Weissenbacher, *Sources of Power: How Energy Forges Human History (Vol. 1, 2)*, Greenwood Publishing Group (2009)

Margaret A. Weitekamp, *Right Stuff, Wrong Sex - America's First Women in Space Program*, Johns Hopkins University Press (2005)

David Whitehouse, *One Small Step - Astronauts In Their Own Words*, Quercus Publishing (2012)

Mark Williamson, *Spacecraft Technology - The Early Years*, Institution of Engineering and Technology (2006)

Web Sources Consulted and Suggested

Russia Beyond
https://it.rbth.com/
https://www.rbth.com/
https://ru.rbth.com/

wikipedia
https://it.wikipedia.org/
https://en.wikipedia.org/
https://ru.wikipedia.org/
https://fr.wikipedia.org/
http://cosesovietiche.blogspot.com/
http://en.roscosmos.ru/
http://it.knowledgr.com/
http://jamesoberg.com/
http://russiaintranslation.com/
http://spaziali.altervista.org/
http://www.ansuitalia.it/
http://www.arar.it/
http://www.astronautix.com/
http://www.copernicus.eu/
http://www.russianspaceweb.com/
http://www.spacefacts.de/
http://www.spacesafetymagazine.com/
https://abcnews.go.com/
https://aerospacecue.it/
https://africanews.space/
https://archivio.corriere.it/
https://argomenti.ilsole24ore.com/
https://articolidiastronomia.com/
https://biografieonline.it/
https://earthsky.org/
https://edition.cnn.com/
https://en-academic.com/
https://eng.mil.ru/
https://global.jaxa.jp/
https://history.nasa.gov/
https://interfax.com/
https://isagitalia.org/
https://it.knowledgr.com/
https://it.sputniknews.com/
https://it.sputniknews.com/person_Valentina_Tereshkova/
https://ita.agromassidayu.com
https://leganerd.com/
https://nssdc.gsfc.nasa.gov/
https://ria.ru/
https://rollingsteel.it/
https://science.jrank.org/
https://scienze.fanpage.it/
https://segretidellastoria.wordpress.com/

https://spacecenter.org/
https://spaceflightnow.com/
https://tass.com/science/
https://thebarentsobserver.com/
https://warspot.ru/18227-nedelinskiy-koshmar
https://web.mit.edu/
https://wsimag.com/
https://www.adnkronos.com/
https://www.aeci.it/
https://www.afp.com/
https://www.agi.it/
https://www.alamy.it/
https://www.ansa.it
https://www.ansa.it/canale_scienza_tecnica/
https://www.ansamed.info/
https://www.archivioluce.com/
https://www.asi.it/
https://www.astronautinews.it/
https://www.astronews.it/
https://www.astroviewer.net/
https://www.bloomberg.com/
https://www.britannica.com/
https://www.chicagotribune.com/
https://www.cicap.org/
https://www.dbpedia.org/
https://www.dire.it/
https://www.eastjournal.net/
https://www.eda.admin.ch/
https://www.esa.int/
https://www.fantascienza.com/
https://www.focus.it/
https://www.fomamag.com/
https://www.france24.com/
https://www.geopolitica.info
https://www.geopop.it/
https://www.globalscience.it/
https://www.globalscience.it/
https://www.ilmessaggero.it/
https://www.ilpost.it/
https://www.ilsole24ore.com/
https://www.independent.co.uk
https://www.internazionale.it/
https://www.issnationallab.org/
https://www.jstor.org/

https://www.kp.ru/
https://www.lastampa.it/
https://www.latimes.com/
https://www.lescienze.it/
https://www.liveinternet.ru/users/adpilot/post375063454
https://www.massa-critica.it/
https://www.media.inaf.it/
https://www.nationalgeographic.com/science/
https://www.nytimes.com/
https://www.passioneastronomia.it/
https://www.passioneastronomia.it/
https://www.planetary.org/
https://www.pravda.sk/
https://www.quora.com/
https://www.rainews.it/archivio-rainews
https://www.reccom.org/
https://www.repubblica.it/
https://www.sciencedirect.com/
https://www.smithsonianmag.com/
https://www.storicang.it/
https://www.sundaytimes.lk/
https://www.telegraph.co.uk/
https://www.theguardian.com/science/
https://www.thetimes.co.uk/
https://www.treccani.it/
https://www.wam.ae/
https://www.washingtonpost.com/
https://www.wikidata.org/
https://www.wired.it/
https://www.youtube.com/watch?v=IHD2X-CvY6g

Printed in the United States
by Baker & Taylor Publisher Services